CORPVS SPECVLORVM ETRVSCORVM

GREAT BRITAIN

2

CAMBRIDGE

CORPUS CHRISTI COLLEGE
THE FITZWILLIAM MUSEUM
THE MUSEUM OF ARCHAEOLOGY AND ANTHROPOLOGY
THE MUSEUM OF CLASSICAL ARCHAEOLOGY

BY

RICHARD V. NICHOLLS

EDITED BY JUDITH SWADDLING AND TOM RASMUSSEN

FITZWILLIAM MUSEUM

CAMBRIDGE UNIVERSITY PRESS

Published with assistance from the
BRITISH ACADEMY

Published by the Press Syndicate of the University of Cambridge
The Pitt Building, Trumpington Street, Cambridge CB2 1RP
40 West 20th Street, New York, NY 10011-4211, USA
10 Stamford Road, Oakleigh, Victoria 3166, Australia

First Published 1993

Printed in Great Britain at the University Press, Cambridge

A catalogue record for this book is available from the British Library
Library of Congress cataloguing in publication data

Nicholls, Richard Vaughan.
Corpus speculorum etruscorum: Great Britain 2 Cambridge / by
Richard Nicholls: edited by Judith Swaddling and Tom Rasmussen.
p. cm
Includes index
ISBN 0-521-43380-0
1. Mirrors. Etruscan–Catalogs. 2. Mirrors–Great Britain–
Catalogs. 3 Italy–Antiquities–Catalogs. I. Swaddling, Judith.
II. Rasmussen, Tom. III Title.
DG223.7.M55N53 1993
739.5'12'09375–dc20 92–35756 CIP

ISBN 0521 43380 0 hardback

CONTENTS

PREFACE

Corpus Speculorum Etruscorum, Great Britain 1 is devoted to the mirrors in the British Museum and at the time of going to press is in preparation; its publication is thus preceded by that of the present slenderer volume covering the Cambridge examples.

The Etruscan and Praenestine mirrors in Cambridge are in four different University institutions, all of which have collaborated readily in allowing their material to be photographed, drawn, recorded and sampled for analysis at the Fitzwilliam Museum. Special thanks are due to the Director and Syndics of the Fitzwilliam and to the Fitzwilliam Museum Publications Committee for sanctioning this work, to the Syndics of the Cambridge University Press for undertaking its publication and to the British Academy for making a grant towards the cost of publication and for helping with editorial expenses, and to all of these for their great patience. The photographs are the work of successive members of the Fitzwilliam Museum's photographic team, the late Frank Kenworthy and Paul Dixon, Andrew Morris and James Butler. Apart from negligible contributions from the author, the facsimile drawings are by Garth Denning. These are meticulous reproductions of the scenes and decoration of the mirrors whereas the modern infill is sometimes inaccurate and misleading. Lines indicating subsequent cracks or other damage have been omitted on those drawings of the reverse of the mirrors where they seem likely to prove unnecessarily confusing. The text was prepared in 1983-5, revised in 1986 and updated in 1987-8, the author having undertaken the task while Senior Keeper and Keeper of the Antiquities at the Fitzwilliam and having finished it following his retirement. He would like to acknowledge the help of the British members of the International Scientific Committee for the *Corpus*, particularly of Dr Glenys Lloyd-Morgan, now retired from the Committee, who helped to institute the project at Cambridge and carefully checked all the drawings, and of Dr Judith Swaddling and Dr Tom Rasmussen, who edited this volume and standardized the English terminology used. The text has also been read by Dr L. Bouke van der Meer and various others who have made valuable suggestions. Help on particular points has been forthcoming from Rachel Edwards, Dr Bettina von Freytag (gen. Löringhoff), Dr Ursula Höckmann and Professor Richard De Puma. The author also wishes to thank many others for their co-operation: Professor R. I. Page, Librarian of Corpus Christi College, Dr D. W. Phillipson and Mary Cra'ster (now retired) of the Museum of Archaeology and Anthropology and Professor A. M. Snodgrass of the Museum of Classical Archaeology, as well as his former colleagues in the Antiquities Department of the Fitzwilliam, Janine Bourriau, Christopher Simon and David Gill, and various past members of the technical staff there, especially Frank Rookes and the late Norman Rayner, and of Julie Dawson and her present team of conservators, in particular Ticca Ogilvie who carried out the recent conservation of the lidded mirror, **3**. Finally, a great burden in seeing this book through publication has fallen on the current Director of the Fitzwilliam, Simon Jervis, and current staff in the Antiquities Department there, particularly Dr Eleni Vassilika, not to mention her husband, Dr Jack Ogden, who has been responsible for the camera-ready copy. The final typescript was prepared by Beverley Fryer, courtesy of the Trustees of the British Museum. Dr J. A. Charles of the University Department of Metallurgy and Materials Science and Barbara Reeve, formerly a conservator in the Antiquities Department of the Fitzwilliam, have given valuable help on questions of technique and condition. The bronze analysis has been carried out by Dr P. T. Craddock and Dr C. Buck through the generous co-operation of the British Museum Research Laboratory, apart from that of **3** (by Dr Jack Ogden).

The format is that prescribed for the series, but with some inevitable economies because of the increased costs resulting from the delay in publication. The entries within the different collections are arranged in inventory order, first by the prefix letters indicating the inventory involved and then by their serial numbers within that inventory. The exception is the uninventoried Lewis collection at Corpus Christi College, where the mirrors are dealt with in the presumed order of their acquisition by their former owner. Some recently agreed improvements to the format are also incorporated. Thus, just as a small black disc on the drawings regularly indicates the drilled centre point of the mirrors, so a small black arrow now marks the area from which the metal sample has been taken for analysis. Also, insets on the plates now provide useful side views of the animal-head terminals of the metal handles. It is regretted that the largest of the Cambridge mirrors have had to be illustrated with their lower parts shown separately; this is simply because they cannot otherwise be fitted into the page-size at the required 1:1 scale. In accordance with the usage of the *Corpus*, figures from Greek mythology are referred to, in the first instance, by the English equivalents of their Greek names, the English spellings used following the system adopted by Sir John Beazley for Attic vases. In the Etruscan entries the Etruscan equivalent is also quoted, if attested, italics being used if the spelling is that of an associated inscription on the same mirror.

The dating of the mirrors remains problematical, with conflicting high and low systems of chronology continuing irreconcilable. The main stages of the stylistic development of their engraving are, however, quite clear and it seems at present better to cite these rather than absolute dates whenever possible. On a moderately low conventional chronology these are: Late Archaic, later 6th–mid 5th centuries BC; Early Classical, mid–late 5th century BC; Classical, late 5th–mid 4th centuries BC; Late Classical, mid–late 4th century BC; Early Hellenistic, end of 4th–early 3rd centuries BC; Hellenistic, 3rd–2nd centuries B.C. These stages must inevitably have overlapped a little, whatever the absolute dating used. These mirrors were enduring pieces of equipment and, where they occur in graves, it is not always clear whether this marks the end of a lifetime's use on the part of their original owner or of one of her descendants. This may also help explain why the grave-group evidence for dating the mirrors is sometimes puzzling.

The drawing following this Preface, by Susan Bird of the British Museum, provides a key to the English terms used for the different parts of the mirrors. Some citations of more recent publications have been added by the author as the editors passed the book to press.

Cambridge, 1992 R. V. Nicholls

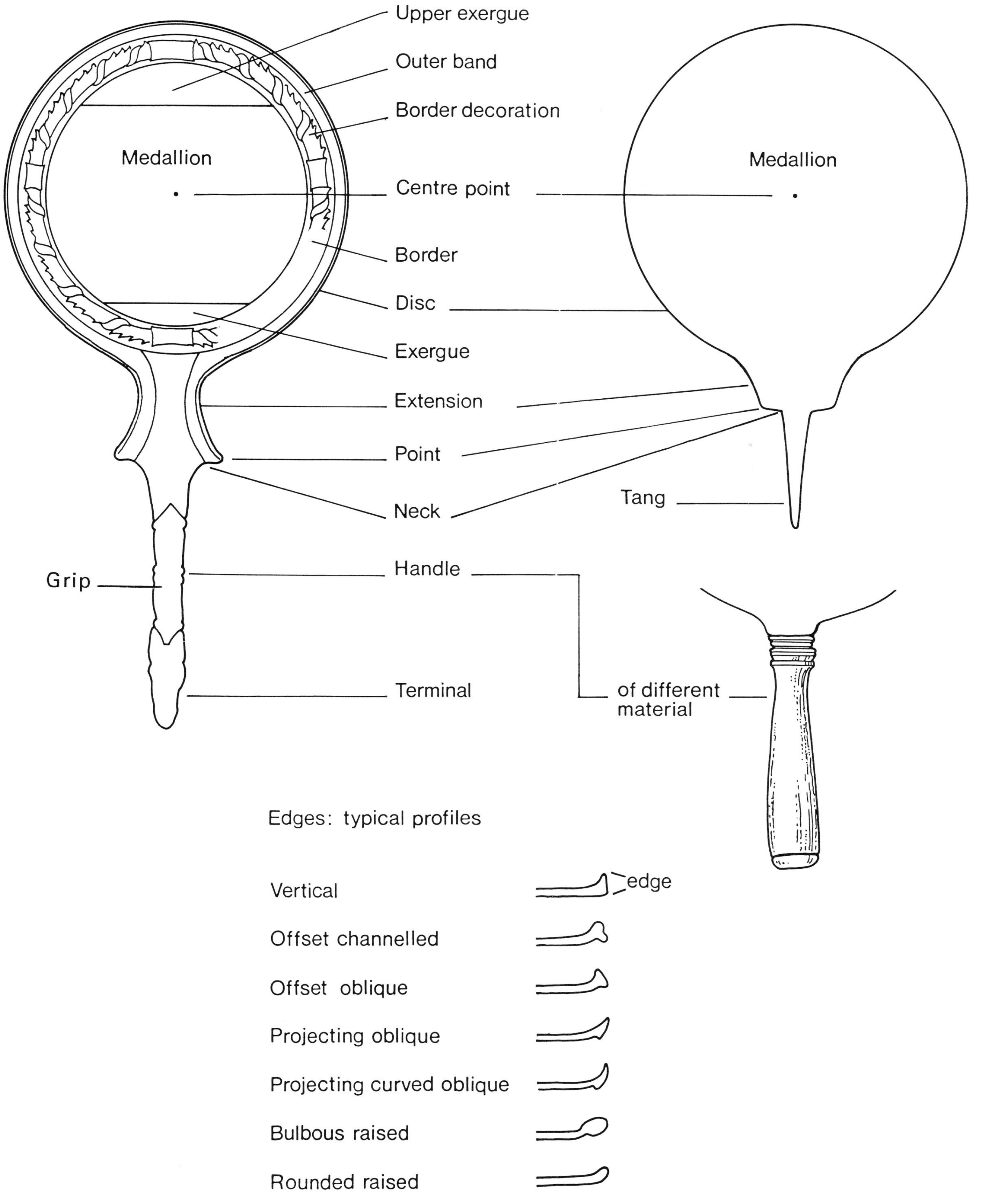

Terms used in describing the different parts of mirrors and their edge profiles.

ABBREVIATIONS

1 Publications

AA — *Archäologischer Anzeiger*, Berlin, since 1889 published with *JdI*, *q.v.*

AbhBerlin — *Abhandlungen der deutschen (preussischen) Akademie der Wissenschaften zu Berlin, Philologisch-historische Klasse*

ActaArch — *Acta Archaeologica*

ADAM, *Miroirs prénestins* — R. ADAM, *Recherches sur les miroirs prénestins*, Paris, 1980

AEArq — *Archivo español de arqueología*

AJ — *The Archaeological Journal*

AJA — *The American Journal of Archaeology*

AM — *Mitteilungen des deutschen archäologischen Instituts, athenische Abteilung*

AnnInst — *Annali dell'Instituto di Corrispondenza Archeologica di Roma*, Rome, 1829-85

AnRepFitzwilliam — *Annual Report of the Fitzwilliam Museum Syndicate*, Cambridge, published initially only in *Cambridge University Reporter*; also in separate format 1910 f.; combined with *Annual Report of the Friends of the Fitzwilliam* (latter issued independently 1909–66) 1967 f.

AntK — *Antike Kunst*

Archaeologia — *Archaeologia or Miscellaneous Tracts Relating to Antiquity Published by the Society of Antiquaries of London*

ArchCl — *Archeologia classica*

ARID — *Analecta Romana Instituti Danici*

Aufnahme fremder Kultureinflüsse — *Die Aufnahme fremder Kultureinflüsse in Etrurien und das Problem des Retardierens in der etruskischen Kunst, Referate vom Symposion des deutschen Archäologen-Verbandes, Mannheim, 8-10.2.1980*, Mannheim, 1981 (*Schriften des deutschen Archäologen-Verbandes* V)

AZ — *Archäologische Zeitung*, Berlin, 1843-86

BdA — *Bollettino d'Arte*

BEAZLEY, *EVP* — J. D. BEAZLEY, *Etruscan Vase-Painting*, Oxford, 1947

BORDENACHE, *Ciste* I, 1 — G. BORDENACHE BATTAGLIA, with collaboration of A. EMILIOZZI, *Le ciste prenestine* I, *Corpus* 1, Rome, 1979

BROMMER, *Denkmälerlisten* — F. BROMMER, with collaboration of A. PASCHLOW- BINDOKAT and D. KEMP-LINDEMANN, *Denkmälerlisten zur griechischen Heldensage* I-III, Marburg, 1971-6

Bronzes hellénistiques et romains — *Les bronzes hellénistiques et romains, tradition et renouveau; actes du Ve Colloque International sur les bronzes antiques, Lausanne, 8-13 mai 1978*, Lausanne, 1979 (*Cahiers d'archéologie romande* no. 17)

BullArchNap — *Bullettino archeologico napoletano*, Naples, 1843-60

BullCom — *Bullettino della Commissione Archeologica Comunale di Roma*

Cat. Castellani, Paris 1884. — *Catalogue des objets d'art... dépendant de la succession Alessandro Castellani* (sale catalogue, Paris, 1884)

CII — A. FABRETTI, *Corpus Inscriptionum Italicarum et Glossarium Italicum*, Turin, 1867

CII, App.	G. F. GAMURRINI, *Appendice al Corpus Inscriptionum Italicarum di A. Fabretti*, Florence, 1880
CIL	*Corpus Inscriptionum Latinarum*
*CIL*²	*Corpus Inscriptionum Latinarum*, 2nd edition
Classical Heritage	R. NICHOLLS, with contributions from G. POLLARD and T. VOLK, *Classical Heritage, Greek and Roman Art from Cambridge College Collections, an exhibition at the Fitzwilliam Museum, Cambridge, 1 August–17 September, 1978*, Cambridge, 1978
ComCAS	*Cambridge Antiquarian Communications, being Papers presented at the Meetings of the Cambridge Antiquarian Society*, Cambridge, 1864-88. Succeeded by *Proceedings and Communications of the Cambridge Antiquarian Society*
COMSTOCK–VERMEULE, *Boston Bronzes*	M. COMSTOCK and C. VERMEULE, *Greek, Etruscan and Roman Bronzes in the Museum of Fine Arts, Boston*, Greenwich, Conn., 1971
CSE	*Corpus Speculorum Etruscorum*, 1981 f.
DE GRUMMOND, *Guide*	N. T. DE GRUMMOND, ed., *A Guide to Etruscan Mirrors*, Tallahassee, 1982
DE SIMONE, *Entleh.*	C. DE SIMONE, *Die griechischen Entlehnungen im Etruskischen* I–II, Wiesbaden, 1968, 1970
DUCATI, *AE*	P. DUCATI, *Storia dell'arte etrusca*, Florence, 1927
EAA	*Enciclopedia dell'arte antica classica e orientale*, Rome, 1958–73
EMILIOZZI, *Collezione Danielli*	A. EMILIOZZI, *La Collezione Rossi Danielli nel Museo Civico di Viterbo*, Rome, 1974 (*Musei e collezioni d'Etruria* I)
FIESEL, *Namen*	E. FIESEL, *Namen des griechischen Mythos im Etruskischen*, Göttingen, 1928 (*Zeitschrift für vergleichende Sprachforschung, Ergänzungsheft* V)
FISCHER-GRAF, *Vulci*	U. FISCHER-GRAF, *Spiegelwerkstätten in Vulci*, Berlin, 1980 (*Archäologische Forschungen* VIII)
FOERST, *GPC*	G. FOERST, *Die Gravierungen der pränestischen Cisten*, Rome, 1978
GazArch	*Gazette archéologique*, Paris, 1875-89
GERHARD, *ES*	E. GERHARD, *Etruskische Spiegel* I–IV, Berlin, 1840–67
GettyMusJ	*The J. Paul Getty Museum Journal*
GIGLIOLI, *AE*	C. Q. GIGLIOLI, *L'arte etrusca*, Milan, 1935
HAYNES, *Bronzes*	S. HAYNES, *Etruscan Bronzes*, London and New York, 1985
HIGGINS, *Jewellery*²	R. A. HIGGINS, *Greek and Roman Jewellery*, 2nd edition, London, 1980
IF	*Indogermanische Forschungen*, Strasburg, 1892-1918
JdI	*Jahrbuch des deutschen archäologischen Instituts*
JHS	*Journal of Hellenic Studies*
KLÜGMANN–KÖRTE, *ES*	A. KLÜGMANN, G. KÖRTE, *Etruskische Spiegel* V, Berlin, 1897
LAMBRECHTS, *Mir. Mus. Royaux*	R. LAMBRECHTS, *Les miroirs étrusques et prénestins des Musées Royaux d'Art et d'Histoire à Bruxelles*, Brussels, 1978
LIMC	*Lexicon Iconographicum Mythologiae Classicae*, Zurich and Munich, 1981 f.
MATTHIES, *PS*	G. MATTHIES, *Die praenestinischen Spiegel*, Strasburg, 1912
MAYER-PROKOP, *Griffspiegel*	I. MAYER-PROKOP, *Die gravierten etruskischen Griffspiegel archaischen Stils*, Heidelberg, 1967 (*RM, Ergänzungsheft* XIII)
MdI	*Mitteilungen des deutschen archäologischen Instituts*, Munich, 1948-53
MedelhavsMusB	*Medelhavsmuseet Bulletin*
MonAL	*Monumenti antichi dell' Accademia Nazionale dei Lincei*
Muse	*Muse, Annual of the Museum of Art and Archaeology, University of Missouri*
NumChron	*The Numismatic Chronicle, Journal of the Royal Numismatic Society*
ÖJh	*Jahreshefte des österreichischen archäologischen Instituts in Wien*
PFISTER-ROESGEN, *Spiegel*	G. PFISTER-ROESGEN, *Die etruskischen Spiegel des 5 Jhs. v. Chr.*, Frankfurt, 1975
PFIFFIG, *Herakles*	A. J. PFIFFIG, *Herakles in der Bildwelt der etruskischen Spiegel*, Graz, 1980
PFIFFIG, *Religio*	A. J. PFIFFIG, *Religio Etrusca*, Graz, 1975
ProcCambrAntSoc	*Proceedings of the Cambridge Antiquarian Society* (see also *ComCAS*)
RA	*Revue archéologique*

RALLO, *Lasa*	A. RALLO, *Lasa, iconografia e esegesi*, Florence, 1974
REA	*Revue des études anciennes*
REBUFFAT, *Miroir*	D. REBUFFAT EMMANUEL, *Le miroir étrusque d'après la collection du Cabinet des Médailles*, Rome, 1973
RepCAS	*Report of the Cambridge Antiquarian Society*, Cambridge, various formats; in 1864-88 issued in same format as *ComCAS*, *q.v.*, but separately paginated
Reporter	*Cambridge University Reporter*
RhMus	*Rheinisches Museum für Philologie*
RICHTER, *MMBronzes*	G. M. A. RICHTER, *The Metropolitan Museum of Art, Greek, Etruscan and Roman Bronzes*, New York, 1915
RM	*Mitteilungen des deutschen archäologischen Instituts, römische Abteilung*
Roma Medio-Repubblicana	*Roma Medio-Repubblicana, aspetti culturali di Roma e del Lazio nei secoli IV e III a.C.*, exhibition catalogue, Rome, 1973
ROSCHER	W. H. ROSCHER, ed., *Ausführliches Lexikon der griechischen und römischen Mythologie*, Leipzig, 1884-1937
StEtr	*Studi etruschi*
StRobinson	*Studies presented to David Moore Robinson on his Seventieth birthday* I-II, St Louis, 1951, 1953
WALTERS, *BMBronzes*	H. B. WALTERS, *Catalogue of the Bronzes, Greek, Roman and Etruscan, in the Department of Greek and Roman Antiquities, the British Museum*, London, 1899
ZAZOFF, *Skarabäen*	P. ZAZOFF, *Etruskische Skarabäen*, Mainz, 1968

2 Other Abbreviations

cm	centimetre(s)
coll.	collection (in Index)
D.	diameter
gm	gramme(s)
H.	height
inv.	inventory number
obv.	obverse, i.e. reflecting side
P.H.	preserved height
P.W.	preserved width
rev.	reverse, i.e. mirror back
rf.	red figure
W.	width
Wt.	weight

I CORPUS CHRISTI COLLEGE

Corpus Christi College was founded in 1352, but its fine collection of classical antiquities was the nineteenth-century creation of one man, Samuel Savage Lewis (1836-1891), Fellow and Librarian of the College and a leading champion of archaeology as a new academic discipline in Cambridge, who bequeathed his collection to the College in 1891. The documentation of the mirrors dealt with here presents some problems, largely because there was no inventory of the Lewis collection, and it will be necessary first to attempt to establish the order in which they were acquired. In 1882 the collection included only two Etruscan mirrors (here **1** and **2**) because, on 6th November 1882, a paper describing them was read to the Cambridge Antiquarian Society (*Reporter*, 1882-3, pp. 161-2; *RepCAS* XLIII, 1882-3, pp. lxxi-ii), its author being Charles William King (1818-1888) of Trinity College, Cambridge, the leading authority on ancient engraved gems. Of these two mirrors, **1** had apparently already been acquired by Lewis by 1879, when it was described by J. de Witt (*GazArch*, 1879, pp. 217-21) as being at the University Museum in Cambridge, possibly signifying that Lewis had lodged it at the Fitzwilliam Museum as a loan exhibit; there seems to be no other information as to just when he had secured **2**. Apparently prompted by the interest generated by King's paper, Lewis bought further mirrors in the years immediately following it; **3** was secured from the Lecuyer sale in Paris in April, 1883, and **4**, **5** and **6** were in the Castellani sale there in May, 1884, although **6** seems to have reached Lewis's hands distinctly later than the others. In 1884 King revised his paper, entitled 'On some Etruscan Mirrors with Engraved Reverses', to include all the new acquisitions as well, apart from **6**, and splendid large facsimile drawings of all five items in time for it to be included in *ComCAS* V, 1880-84, as pp. 185-200; the final part of this volume was not published until 1886, but the fascicule covering the meetings of 1882-3 appeared before the end of 1884. An important source of documentation for the other Lewis antiquities is to be found in the stickers attached to them, minutely inscribed in Lewis's own hand, but these have adhered badly to metal and, in the case of the material of concern here, only survive adequately on **3**. All of the Lewis mirrors were included in the exhibition, *Classical Heritage, Greek and Roman Art from Cambridge College Collections*, at the Fitzwilliam Museum, 1st August - 17th September, 1978. In 1991 the Master and Fellows of Corpus Christi College lodged the whole Lewis collection of antiquities, including the mirrors, on ten-year loan at the Fitzwilliam Museum.

1. Engraved mirror. Figs. 1a-d

Uninventoried. From Cortona. Bequeathed in 1891 by S. S. Lewis, who had apparently already acquired it by 1879; formerly collection of P. A. Laparelli. (Fitzwilliam Museum loan register Ant.1991/103.35.)

CII, no. 1022*bis*, glossary p. 2094; S. BUGGE in *Jenäer Literaturzeitung* II, 1875, p. 287; *Academy* XIV, 1878, p. 227; J. DE WITTE in *GazArch*, 1879, pp. 217-21, with fig.; I. TAYLOR, letter in *Athenaeum*, 20 Sept. 1879, and in *Academy* XVI, 1879, p. 214; A. S. MURRAY in *Academy* XVI, 1879, pp. 195, 214;'A.B.' in *RA* Ser. 2, XXXIX, 1880, p. 114; *Reporter*, 1882-3, p. 162; *RepCAS* XLIII, 1882-3, p. lxxi; S. BUGGE, *Etruskische Forschungen und Studien* IV, Stuttgart, 1883, pp. 29-30; B. LEWIS in *AJ* XL, 1883, p. 36; C. W. KING in *ComCAS* V, 1880-4, pp. 190-2, pl. 4; J. MARTHA, *L'art étrusque*, Paris, 1889, p. 551; KLÜGMANN – KÖRTE, *ES*, pp. 65-6 no. 51*a*, with text fig.; C. PAULI in ROSCHER, III, 1, pp. 1254-5, *s.v.*'Pakste'; G. HERBIG in *IF* XXXVII, 1916-17, pp. 178-9; J. BAYET, *Herclé, étude critique des principaux monuments relatifs à l'Hercule étrusque*, Paris, 1926, pp. 192-4; FIESEL, *Namen*, pp. 55, 87 n. 303; DE SIMONE, *Entleh.*, I, pp. 72 no. 15, 96, II, pp. 12, 33, 34, 60, 97, 128, 145, 168, 172, 180, 196 n. 136, 299, 310, 317, 327; M. LAUNEY, *Études thasiennes* I, Paris, 1944, p. 152; T. J. DUNBABIN (initials misquoted) in *StRobinson* II, p. 1172, pl. 93*a*; *Classical Heritage*, p. 15 no. 56; S. J. SCHWARZ in *LIMC* V, p. 232 no. 332, *s.v.*'Herakles/Hercle'.

Bronze. Lower part of tang broken away. Manually cleaned, but retaining smooth patina, partly deep green malachite, partly greenish black. Considerably pitted. Barely visible vestiges of centre point on rev., near buttock of engraved rider. Faint traces of earlier modern (or original?) granular whitish infill at few points in engraving; more abundant evidence of modern French-chalk infill. On technique, see Appendix.

D. 15.8 cm. P.H. 17.9 cm. Total H. of part of mirror apparently exposed above handle (see below) 16.6 cm. P.H. of tang 2.3 cm. Wt. 225.2 g. Composition (see p. 54): copper 93.5%, tin 4.15%.

Circular mirror with tapered tang. The way in which the engraved decoration of the *obv.* extends on to its upper part and the different patination of its once-socketed section show that the upper, outward-curving part of the tang was not embedded in the handle and remained proud as a rudimentary extension of the disc.

Obv. Slightly convex reflecting face bounded by plain moulding of semicircular section. Projecting channelled edge with relief decoration of tongue pattern with raised outlines. Top of tang and adjoining lower part of disc engraved with a pattern of volutes and palmettes, much of which has been obliterated by repeated polishing during the mirror's ancient period of use and what survives is now quite faint.

Rev. Slightly concave. An U-shaped line above the tang develops to either side into 'running-dog' patterns encircling the medallion and apparently representing waves; these meet at the top as a pair of volutes flanking an inverted nine-petalled palmette and separated from it by a cross-hatched band. Within this border is engraved a rearing stallion apparently named as *Pakste* (probably Pegasos) in an accompanying inscription (see below) and intended as flying through the air towards the spectator's left with his ears back, teeth bared and mane and tail apparently intended by the artist as streaming behind. He is harnessed and has a cord with two *bullae* around the top of his neck which is also encircled lower down by a string of *phalerae*. Riding astride the horse is a disproportionately large beardless young man with close-cropped hair indicated by short, curving lines, named in an accompanying inscription (see below) as *Herkle* (Herakles), although he carries none of that hero's usual attributes. He is naked apart from a *chlamys* looped over his left upper arm and flapping behind and he holds the reins in the left hand while his right touches the horse's mane behind its head, perhaps in a calming gesture. Behind the horse is a dolphin that has leapt from the waves and is at the point of re-entering them. This creature forms an integral part of the medallion composition, but its main function is probably to indicate that the encircling 'running-dog' pattern represents the waves of the sea and that horse and rider are shown as flying over the water.

Inscriptions. Behind the rider's shoulder, written from right to left (except for the letter, *k*, which faces the other way both times) in two slightly overlapping lines:

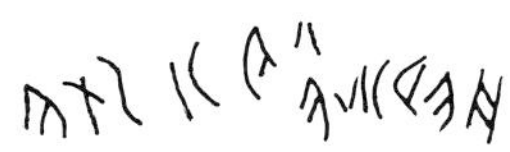

Pakste
Herkle

Although BUGGE attempted to derive *Pakste* from Παγασίτης and treat it as an ethnic (i.e.'Herakles of Pegasai'), without explaining the relevance of the scene to Herakles' interlude with the Argonauts, it has otherwise generally been accepted as an Etruscan transcription of Pegasos and as a variant of the rather more usual *Pecse*. In DE SIMONE, *Entleh.* I, p. 72 no. 15, *Herkle* has been misread as *Hercle*, but there can be no doubt as to the spelling. The developed form of the *r* and *t* may argue against too early a date and the late retention of the k-spellings and the form of the *e* may suggest a northern origin.

Subject. Herakles riding over the sea on a wingless flying horse, probably Areion, but named as *Pakste* (Pegasos?). The possible explanation (*Classical Heritage*, p. 15) that the engraver has confused Herakles and Bellerophon seems disproved by a number of related Praenestine scenes clearly depicting Herakles; see **5**, especially p. 21. DE WITTE and others explained the presence of the sea and dolphin by wrongly equating Herakles–Melqart with Melikertes. In fact, both waves and dolphin have a special aptness, since Areion's father was the sea-god, Poseidon

(Nethuns). For the identification of the horse as Areion, see especially DUNBABIN, *loc. cit.* Famous horses are occasionally misnamed on other mirrors as well. Thus, on an Etruscan mirror in Paris (REBUFFAT, *Miroir* no. 51), the wooden horse of Troy is named *Pecse* (Pegasos), while on a Praenestine mirror in London, British Museum GR 1897.10-30.2 (ADAM, *Miroirs prénestins* no. 3; I. KRAUSKOPF in *LIMC* II, p. 478 no. 5, pl. 358) showing a conversation between Oinomaos (*Oinomavos*) and Bellerophon (*Melerpanta*), the winged steed between them is called not Pegasos but Areion (*Ario*).

Engraving Characteristics. The cutting is of moderate depth and reasonably firm, although there are occasional breaks and awkward overlaps, e.g. on the left leg of Herakles, above the knee, and these are especially frequent on the 'running dog' wave pattern. The dolphin shows vestigial traces of the preliminary cutting of a somewhat different outline under its head and above its tail. Extensive use is also made of short incisions, e.g. on Herakles' hair (cf. PFISTER-ROESGEN, *Spiegel*, nos. S8, S17, S28 and the later Praenestine mirror, *ibid.* no. S59; also, here, the much later **2**). The eye in his profile head is shown full-face and leaf-shaped. Both horse and dolphin are given eyes and eyebrows of human type, again shown full-face in profile heads. Short lines indicate the chest, shoulder, rib-cage and leg musculature of the horse, the folds of skin at his joints, his mane and tufts of hair at his fetlocks and over his hooves. The fins and tail of the dolphin are given fish-like scalloped edges and spiny ribs. The 'running-dog' wave pattern constitutes an early instance of the 'encompassing groundline' or 'encircling landscape' in Etruscan engraving (N. T. DE GRUMMOND in *AntK* XXV, 1982, pp. 3-14). The same motif is used this way even earlier on mirrors in Milwaukee and Boston, in the latter already moved inside the main border; both of these apparently show Helios (Cath, Cautha or Usil) rising above the waters of Okeanos: MAYER-PROKOP, *Griffspiegel*, nos. S2, S3, pl. 3; COMSTOCK-VERMEULE, *Boston Bronzes*, no. 391 (dated too late); *CSE* USA, 1 no. 26; I. KRAUSKOPF, *LIMC* V, pp. 1040-1, nos. 1, 12, pls. 654-5, *s.v.*'Helios/Usil'. Later, on a 4th-century BC Vulci mirror in London, British Museum GR 1966.3-28.13, showing Peleus and Thetis (Haynes, *Bronzes*, no. 177), the encircling wave pattern is completely absorbed into the medallion scene.

Fabric and Date. Etruscan, possibly rather provincial north Etruscan, Early Classical, but apparently influenced by earlier mirrors. A certain clumsiness in design, proportions and execution seems to suggest a provincial origin, as may also the rather low tin content of the metal (cf. p. 000). The mirror was found at Cortona and oddities of spelling and letter forms suggest that it may also have been made in northern Etruria. Echoes of early features include the form of the 'running-dog' border and the way in which the profile heads are engraved with full-face eyes. That the execution is, however, later seems to be shown by the Free-Style rendering of the flapping *chlamys*, by the treatment of the animal heads, by certain of the letter forms and by the character of the *obv.* florals.

2. Engraved mirror. Figs. 2a-d

Uninventoried. Unprovenanced. Bequeathed in 1891 by S. S. Lewis, who had already acquired it by 1882; apparently previously in the collection of Prince Barberini (this earlier documentation rests on the identification, already made by KING, of GERHARD, *ES*, pl. 380, 1 as a faulty and incomplete drawing of its medallion scene with the border omitted and an arbitrary mirror-outline added; this must indeed be the case, despite its inappropriate description there as *stark verletzter*, unless it has an exact double of unprecedented form). (Fitzwilliam Museum loan register Ant.1991/103.38.)

GERHARD, *ES* IV, Part V, pp. 27-8, pl. 380, 1; *Reporter*, 1882-3, pp. 161-2; *RepCAS* XLIII, 1882-3, pp. lxxi-ii; C. W. KING in *ComCAS* V, 1880-4, pp. 188-90, pl. 2; *Classical Heritage*, p. 15 no. 59.

Bronze. Heavily cleaned manually, but now presenting a blotchy surface with much of *rev.*, handle and lower part of *obv.* retaining thin brown layer of vestigial oxide, but with patches of bare bright metal (whether or not originally so preserved by a tin oxide patina it is now difficult to tell), including a diagonal band across the *rev.*, and with more limited areas of heavier corrosion showing green malachite over localised cuprite. Centre point on *rev.* at edge of engraved column. Clumsy and occasionally inaccurate modern white-ink infill on *rev.* engraving. On technique, see Appendix.

D. 17.55 cm. H. 32.7 cm. H. of disc and extension 20.5 cm. H. of handle 12.2 cm. W. at extension points 4.1 cm. Wt. 565.95 gm. Composition (see p. 54): copper 89%, tin 10.51%.

Disc almost circular with concave-sided extension with slightly projecting points; handle cast in one piece with disc and ending in terminal in form of stylized head of hind (or fawn?).

Obv. Slightly convex reflecting face bounded by a shallow relief band which continues around the extension and the neck and grip of the handle. Projecting oblique edge decorated with pattern of tongues with raised outlines. Grip and neck of handle carry cast relief decoration in form of an elongated floral(?) motif which continues on to lower part of extension as scallop-like frill between side-petals(?). Above, on a curved base-line running parallel with the top of this, the extension is engraved with a winged griffin shown reserved against a stippled background. It has a lion's body and an eagle's beak, with part of a leonine mane appearing below the beak, but, apart from faint traces of part of its back and one wing, all of the middle part of this figure has been completely worn away by the repeated ancient polishing of the reflecting face.

Rev. Handle ground flat, apart from hollows behind terminal. Disc and extension somewhat concave and carrying engraved decoration with double lines framing the medallion, exergue and extension zones. That of the extension is occupied by the three-quarter bust of a woman with a necklet rendered by dots. She wears her hair combed back in a style like that of the goddess on the medallion (see below) and may possibly, like her, be intended as Aphrodite (Venus). The two sprays of the garland border emerge from behind the bust and consist of alternating pairs of horizontal and oblique leaves. Garlands of this type are conventionally identified as of laurel (cf. R. PAMPANINI in *StEtr* IV, 1930, pp. 293-320), but might alternatively be a stylization of myrtle. The sprays terminate in rather inappropriate trumpet-like flowers flanking a central rosette. The exergue shows a stippled semicircular mound(?) from which emerges a seven-petalled palmette. Its petals have centre-ribs and are interspersed by single berries(?) on long stalks, and by fleshy-looking flowers, one of them with a stippled centre, also on similar stalks, but carrying isolated small leaves as well. From behind the flowers emerge stalks bearing three berries(?).

The medallion shows a three-figure scene. To the right is a woman seated on a solid throne with ornate volute decoration, perhaps an outdoor stone ceremonial throne, with outcrops of rock in front of it shaped to fit each of her feet in lieu of a footstool. She wears shoes, a sleeveless *chiton* fastened at the shoulders and a *himation* draped across her lap and legs and around her left shoulder and back, with its edge held in her right hand, possibly in an unveiling gesture. Her left hand rests in her lap and she wears a snake bracelet on her right wrist and an earring with a round-tipped tapered pendant in her left ear. Beside her is a free-standing column on top of which perches a dove (misinterpreted as a raven by GERHARD and KING). The column has a plain abacus and echinus and carries horizontal bands of decoration with zigzags and rows of circles. Both column and dove may be attributes of the enthroned woman. As she is depicted on a larger scale than the other figures she is likely to be a goddess and, on the evidence of her iconographic type (e.g. cf. the earlier Etruscan example, PFISTER-ROESGEN, *Spiegel*, no. S58) and of her dove attribute, most probably Aphrodite (Venus). Facing her is a man seated on a similar but smaller solid throne, also decorated with volutes but lacking any foot supports. He is short-haired and beardless, apart from sideburns. His legs and hips are wrapped in a *himation* and he seems to be in conversation with the goddess, gesticulating with his right hand while his left rests in his lap. Beside him stands a woman, also looking towards the goddess with her right arm round his neck and her left akimbo, the hand resting, palm outwards, against her hip. She is wearing an ungirdled sleeved *chiton*-like garment which seems to have a stripe around its neck and may thus be a *stola*; there is also a semblance of vertical stripes, but these may be unintentional and due to a clumsy rendering of the drapery folds. The woman's frizzy hair protrudes to the back, where it is bound by a band. She has a snake bracelet on her left wrist and wears an earring, apparently with three supports to a single pendant, and two necklets, the upper rendered as a row of dots, the lower as a band with pendants.

Subject. A goddess, probably Aphrodite (Venus), with a man and young woman in a conversation group. Described by GERHARD as Aphrodite with Paris and Helen and by KING, doubtless misled by his identification of the bird as a raven, as a group at an oracle. A mirror by the same engraver from Tusculum with an analogous scene (see below) may at first seem to support GERHARD's view that the couple are a pair of lovers before the goddess, but, whereas he could be right in identifying the young woman as Helen, the man occupies an elaborate throne matching that of the goddess. Is he one of her paramours, Adonis or perhaps, to judge from the way in which the woman has embraced him, Phaon?

Engraving Characteristics. The cutting was originally reasonably firm, but shows distinct shortcomings in draughtsmanship, as, e.g., in the awkwardly foreshortened feet of the seated man or the curiously elongated treatment of the standing woman's left hand. The profile eyes are competently shown in an advanced style, usually with over-and under-lids, and the three-quarter face of the *rev.* extension is also competently if summarily rendered. Short lines suggest the sternum, pectoral, thorax and *linea alba* of the seated man and the flesh folds at his navel. But there is a distinct awkwardness in the proportions and structure of the figures and an element of mannerism, e.g. in the gestures of the rather large hands, as with the goddess's separating the first finger from the others and the man's keeping the little finger and its neighbour pressed together. The trumpet-like flowers at the top of the garland are shown in perspective.

Fabric and Date. Praenestine, Late Classical. The disc is still of early, broadly circular shape, but the perspective rendering of the trumpet-shaped flowers of the border and the three-quarter treatment of the extension bust echo developments in Apulian vase-painting of the third and last quarters of the 4th century BC; for such busts and *protomai* on Praenestine mirrors, cf. ADAM, *Miroirs prénestins*, pp. 59-61, fig. 5. For the griffin of the *obv.* cf. BORDENACHE, *Ciste* I, 1, pls. 113-17, 159-64; the stippled background recurs with Late Classical Etruscan examples, e.g. GERHARD, *ES*, pls. 165, 167, 217. The detached nostril-line of the medallion figures is a debased form of a device characteristic of the finest 4th-century Praenestine engraving, as on the Ficorini (T. DOHRN, *Die Ficorinische Ciste in der Villa Giulia in Rome*, 1972, pls. 1, 4-15) and Chrysippos *cistae* (GIGLIOLI, *AE*, pl. 292; FOERST, *GPC*, no. 95) in the Villa Giulia and some early mirrors (e.g. HAYNES,

Bronzes, no. 17), while the goddess's earring seems near the Hellenistic Period (cf. HIGGINS, *Jewellery*^{2}, p. 151, pl. 42*a*). The best evidence for dating, however, is provided by a mirror with an analogous scene apparently by the same hand from the Tomb of the Rabirii at Tusculum, where the finds have been dated between the late 4th and the 3rd centuries BC: M. BORDA in *Bullettino del Museo della Civiltà Romana* XIX, pp. 31-2, pl. 2 (Suppl. to *BullCom* LXXVI, 1956-8); *Roma Medio-Repubblicana*, pp. 305-11 no. 453, pl. 105. The artist may be termed the Loving Couples Master; near him is a mirror in Stockholm also with a griffin on the *obv.*: I. WIMAN in *MedelhavsMusB* XVII, 1982, pp. 37-41, figs. 1-3.

3. Lidded mirror with *appliqué* relief. Figs. 3 a-d

Uninventoried. Apparently from Corinth. Bequeathed in 1891 by S. S. Lewis, who had acquired it in 1883; previously Lecuyer collection. Carries sticker (temporarily removed during photography) in Lewis's hand:'From Corinth. Lecuyer collection, no. 444: Paris: 370 frcs. Hoffmann.' As provenance (not quoted in Lecuyer catalogue) is given without qualification by Lewis, who was highly scrupulous in such matters, it may be taken that he had it on good authority. Wrongly described by KING as acquired from the Charvet collection. (Fitzwilliam Museum loan register Ant.1991/103.43.)

Terres cuites de Tanagra et d'Asie Mineure, collection Camille Lecuyer (sale catalogue, Paris, 1883) p. 70 no. 444; C. W. KING in *ComCAS* V, 1880-4, pp. 187-8, pl. 1; WALTERS, *BMBronzes*, p. 127 under no. 732; C. A. DI STEFANO in *ArchCl* XXII, 1970, p. 80, pl. 27, 2; *Classical Heritage*, pp. 16-17 no. 64, with fig.

Bronze, with an infill of a different metal that shows opaque on X-rays behind much of the bodies of the figures of the *appliqué* relief, recently revealed as a layer of ancient lead behind the man in the middle, but behind the damaged parts of the right-hand figure a modern filler containing barium salts (mostly replaced in 1991) gave a similar opaque effect on X-rays. Mirror and lid, formerly corroded together, separated in 1991. Missing are parts of flanges of both, outer portions of hinge and all of wire pull for opening mirror, apart from one of its fixings and faint traces of where the other was. Bronze shows dull green malachite patina with some areas of bright green, particularly on the *rev.* and *obv.* of the mirror, of cuprite, particularly on the interior of the lid, and patches of bright metal with a tin oxide patina on the mirror face and lid interior; some manual cleaning, especially over the lid reliefs, exposing cuprite. At some time after its publication by KING in 1884 (when shown repaired but largely intact) and before the 1960s, relief suffered damage to lower right part, losing sections of ground on which figures stand, including other pull attachment, and parts of legs and drapery of Muse. This damage had been made good with a modern filler covered with a spurious malachite patina and with some of the surviving fragments wrongly located. To preserve the mirror, this restoration had to be replaced in the Fitzwilliam Museum in 1991.

D. 13.8 cm. P.H., including vestiges of hinge and rivets, 14.3 cm. Depth when closed 2.5 cm. Wt. (before 1991 restoration) 728 gm. Bronze surfaces accessible in 1985 completely corroded, with no sound metal left, and thus unable to be analysed. Wavelength dispersive X-ray fluorescence microprobe analysis carried out by Dr Jack Ogden in 1992 gave the following results: cover interior, copper 85.7%, tin 14.2%; mirror face, copper 92.7%, tin 7.05%.

Circular mirror with hinged lid, both cast as flanged discs and wheel- or lathe-finished.

Mirror obv. Slightly convex reflecting face; wheel-trimmed(?) bands on flange, the projecting uppermost one providing surface on to which lid closed.

Mirror rev. Wheel-cut(?) relief pattern of wide and narrow raised concentric circles around a shallow centre point and continuing out to the flange.

Lid rev. (i.e. inner surface). Almost flat with similar but shallower pattern of wide and narrow raised wheel-trimmed concentric circles extending out to the angle with the flange.

Lid obv. Apparently slightly convex face bounded at edge by raised band of hemispherical section designed to conceal edge of hammered circular sheet-bronze *appliqué* relief soldered within it. Wheel-trimmed(?) bands on flange. Outer edge of hammered relief disc is occupied all round by engraved scallop-like pattern. At the bottom of the disc these scallops are much taller and more rock-like, filling a sort of exergue under the relief base-line of the main scene, and at the sides of the disc their interstices are sometimes occupied by projecting lines, possibly suggesting that they are intended as rocks with blades of grass growing between them; i.e. the border pattern may be a distinctly stylized form of the'encircling landscape' device, for which see p. 15. At the top of the relief disc, where it is soldered over the remains of the folded-over sheet of bronze that formed one of the hinge-plates, is an engraved floral decoration. To the right of the head of Dionysos (see below) there is a lotus blossom and, from the left of this, an undulating stem leads to a smaller lotus flower above his head. The same pattern repeated itself to the right, somewhat displaced by the tip of Eros' left wing, but is not so well preserved.

Three figures are shown in relief moving towards the spectator's right over ground sprouting large daisy-like flowers. All three are garlanded with ivy bearing clusters of berries and their garlands are bound with fillets, loose loops of which hang down on to their brows. The procession is led by a young woman, apparently a Muse, dressed in a sleeveless girdled *chiton* pinned at the shoulders and a *himation* looped over her right forearm and apparently carried around her back and over her concealed left arm. Her *chiton* is being blown back against her legs and the end of the *himation* under her right arm is also caught in the wind. She has her weight on her advanced left leg and her head turned to her right and, with a plectrum in her right hand, is playing a *kithara* held against her left shoulder. She is followed by a young man who is unmistakably the god Dionysos. He is naked save for a *himation* with weighted corners looped over his left arm and also apparently caught in the wind as it trails along the ground behind him. He has a ribboned *thyrsos* carried across his shoulders and gripped in his left hand. His weight is borne on his extended right leg, but he is in fact leaning back tipsily, completely off-balance and supported by an adjoining figure of Eros, who has his left shoulder under Dionysos' right armpit. The latter looks down fondly at Eros who returns his gaze. Eros is nude and striding forward with his left leg. With his right hand he grips Dionysos' right wrist and his left hand seems to be resting on the other god's right shoulder. He, too, may be intended as tipsy, with his right wing furled and his left half-spread.

Subject. Dionysos (Fufluns) with Eros and a Muse. C. A. DI STEFANO has assembled sixteen lidded mirrors of the same type with reliefs showing the same scene (*ArchCl* XXII, 1970, pp. 76-91, pls. 22-31); of these, in her judgment, the present example stands out from the remainder 'per la morbidezza del modellato, per il trattamento dei panneggi e per la finezza dell'esecuzione'. See also LAMBRECHTS, *Mir. Mus. Royaux* no. 14; M. L. ANDERSON in *GettyMusJ* IX, 1981, pp. 59-62, figs. 2-3; H. SALSKOV ROBERTS, *CSE* Denmark 1, no. 8; E. H. RICHARDSON in DE GRUMMOND, *Guide*, pp. 17-19, fig. 18; D. WILLERS in *GettyMusJ* XIV, 1985, pp. 20-6; I. JUCKER in *RM* XCV, 1988, p. 36, pl. 2, 2; E. POCHMARSKI in *AA*, 1988, pp. 487-99, figs. 1-10; B. v. FREYTAG-LÖRINGHOFF, *CSE* W. Germany 3, nos. 6, 20. DI STEFANO has drawn attention to the possible influence on the composition of Greek sculptures of about the later 4th century BC. Thus the Eros and Dionysos form one of the variants of a group of the drunken Dionysos similarly supported by a satyr and the kitharist seems loosely inspired by a sculpture of a Muse, possibly Erato. Eros normally belongs to the circle of Aphrodite and the Muses to that of Apollo and neither are obvious members of the *thiasos* of Dionysos. Their presence here may, however, be symbolic of the combination of love, music and wine. All are garlanded for the Bacchanalia, a suitable subject for a lady's mirror because of the leading role played by women in these rites.

Style Characteristics. The figures are rendered in fairly high relief and with fine, sensitive detail. Although they were doubtless originally formed by hammering in a die, they have subsequently been reworked very extensively in the *repoussé* technique and channelling from the front has also been used to accentuate the outlines. This technique is particularly obvious at the sides of the faces of Dionysos and the Muse that are closest to the relief ground and on the drapery. A circular punch seems to have been used for Dionysos' left nipple and for the petals of the flowers on the ground and a small pointed one on the fillets and on the shaft and ribbon of the *thyrsos*. Apart from the encircling rocky-landscape(?) border and the floral ornament by the hinge, incision has also been used for the feather detail of Eros' wings, the fatty fold on the abdomens of the male figures and the decoration of the *kithara*.

Fabric and Date. Etruscan, Hellenistic. Mirrors of this kind seem to have been produced in Etruria from the late 4th century BC on, but, in the present case, as already noted, the composition is of Early Hellenistic Greek inspiration and, besides that, the figures themselves, particularly that of Dionysos, already show Hellenistic proportions. The mirror is, however, quite certainly an Etruscan product and the suggestion in *Classical Heritage*, based on its unusual technique, that it was a Greek copy of an Etruscan type certainly cannot be sustained. Nevertheless, its provenance from Corinth is of considerable interest, particularly as a second example of the same kind and showing the same scene, now in Marseilles, is cited as from Greece proper and as probably from Corinth (W. FROEHNER, *Musée de Marseilles, catalogue des antiquités grecques et romaines*, Paris, 1897, p. 160 no. 793; DI STEFANO in *ArchCl* XXII, p. 78, pl. 24, 2). Two different hypotheses might explain their presence there, either (a) that they were Etruscan exports incorporated in burials at Corinth at some date before that city's destruction in 146 BC, although the abandonment of the city after that was not total, or (b) that they were heirlooms preserved by their Italian owners for centuries and brought by them to Corinth after that city's refoundation in 44 BC as a colony for Roman veterans and freedmen.

4. Engraved mirror. Figs. 4a-d

Uninventoried. Unprovenanced. Bequeathed in 1891 by S. S. Lewis, who had acquired it in 1884; previously Castellani collection. (Fitzwilliam Museum loan register Ant.1991/103.39.)

Cat. Castellani, Paris, 1884, p. 48 no. 427; C. W. KING in *ComCAS* V, 1880-4, pp. 192-4, pl. 6; *Classical Heritage* p. 16 no. 60.

Bronze. Upper part of disc cracked with two small fragments missing. Green malachite patina over localised cuprite, corrosion being severe over handle and parts of *obv.*; manual cleaning over *obv.* extension and *rev.* disc and extension. Modern pink infill in *rev.* engraving. Faint vestiges of centre point just detectable on *rev.* by right hip of figure of Aphrodite. On technique see Appendix.

D. 14.9 cm. H. 30.4 cm. H. of disc and extension 17.6 cm. H. of handle 13.0 cm. W. at points 3.7 cm. Wt. 285.2 gm. Composition (see p. 54): copper 85.5%, tin 8.97%.

Circular disc with concave-sided extension projecting moderately at points; handle cast in one piece with disc and ending in terminal in form of stylized head of hind (or fawn?).

Obv. Moderately convex reflecting face bounded by narrow channel and rounded moulding, both of which continue around the extension and, to a lesser degree, along the side of the grip of the handle, which is otherwise of slightly convex section. Plain edge of offset curved oblique section. Extension preserves faint vestiges of engraved decoration in form of pair of double volutes of which lower two flank and enclose a palmette and upper two originally supported a crowning palmette and possibly had half palmettes outside them. Its poor preservation seems due partly to heavy oxidation, more to repeated ancient polishing of the reflecting surface.

Rev. Disc has slightly sunken concave medallion surrounded by flat raised border zone, whose plane continued down through extension. Relief decoration of handle grip now difficult to interpret because of heavy corrosion; possibly extremely elongated floral motif intended, with the round shape engraved towards neck of handle perhaps intended as stamen of flower. V-shaped hollow under animal-head terminal. The extension is engraved with a plump-bellied winged boy, probably Eros, wearing shoes and holding a cord, possibly a snare, leading to a dove perched on the border garland. Two rather schematic sprays of leaves, probably intended as laurel, emerge from behind the head and wings of Eros and form a garland occupying the border zone; between them at the top there is a four-petalled rosette. A double horizontal line separates the medallion scene from the Eros. Above this there is a band, half hatched obliquely to the left and half obliquely to the right. The medallion figures are shown in front of part of the colonnade of a building, of which five fluted columns are indicated with 'Aeolic' capitals and a length of the epistyle. The capitals are of a type with necking, squat volutes and no abacus, as often shown on 'raised-border' and Spiky Garland mirrors (cf. A. CIASCA, *Il capitello detto eolico in Etruria*, Florence, 1962, pls. 20,1, 23,2, 24,2, 26,1-3, 28,1, 29,2-3). To the left stands a young man named in the accompanying marginal inscription (see below) as *Elchsuntre* (Alexander, i.e. Paris). He faces the other figures and stands with his left leg crossed over his right and his right hand resting on his club. He wears a Phrygian cap with the flaps hanging down, a girdled short *chiton* pinned at the shoulders and with its lower border indicated, a similarly bordered *chlamys* looped over his left forearm and skin boots bound with thongs around his lower calves. On his right arm is a cross-hatched archer's guard and his left forearm and hand are extended perhaps in a gesture of hesitant choice. Three goddesses face him, all in a similar posture, standing with their weight on their right legs and with their left arms lowered and their right hands raised in the same gesture; they all have similar hairstyles and wear similar shoes, necklets with round beads or pearls and earrings with pendants. The first of the three is named in the marginal inscription as *Uni* (the goddess, Hera). She wears a *peplos* with a long girdled overfold and with bordered edges. The next figure, standing slightly further to the front and so partly overlapping her, is named in the marginal inscription as *Turan* (Aphrodite). Her garment is probably a similar *peplos*, but unfastened so that it leaves most of the front of her body bare, with a corner caught between her legs, and part pulled up over the back of her head and around her shoulders as a veil. The last goddess is bent slightly forward to fit the curve of the medallion and named in the margin as *Menrva* (Athena). Her right hand, while raised in the same gesture, also lightly clasps a spear and she is wearing a similar bordered *peplos* with a long girdled overfold and with its open side at her left and an oblique scalloped-edged aegis over it across her left shoulder. The aegis is adorned with a small Gorgoneion and two four-pointed rosette-stars (possibly symbolic of Athena's half-brothers, the Dioskouroi?). At the top of the medallion, either appearing above the colonnaded building as part of the scene or else forming a separate upper exergue distinct from it, is the head and shoulders of a winged female spirit facing to the spectator's left. She is depicted with her wings spread, on a smaller scale than the figures immediately below but wearing a hairstyle and earring similar to theirs.

Inscriptions. Written from right to left in the margin of the medallion beside Paris' right arm and beginning above his right shoulder:

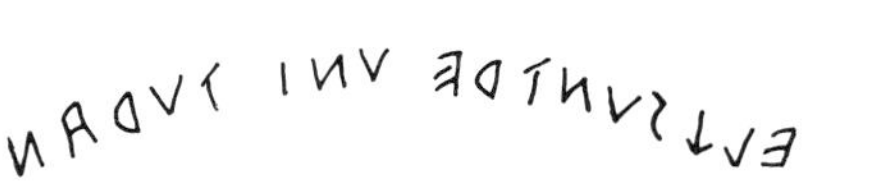

Elχsuntre Uni Turan

There are spaces between the names, but no word punctuation. For the spelling of *Elχsuntre*, with an initial *E* instead of *A*, see DE SIMONE, *Entleh.*, pp. 56-7, for the present spelling, *ibid.* nos. 5, 6, 12, 13. Written from right to left in the margin of the medallion beside Athena's left arm:

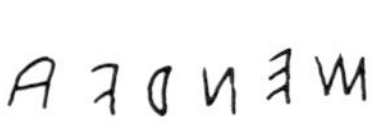

Menrva

Subject. The judgement of Paris. Most probably the winged figure above is either a Lasa or an Etruscan goddess such as Achvizr or Thalna in attendance on Turan (Aphrodite), while Eros and her dove are represented below. As not infrequently in Etruscan art, the contest between the goddesses is shown not in the wilderness but in an elaborate architectural setting.

Engraving Characteristics. The *rev.* engraving is of moderate depth and evenly and precisely cut. The profile eyes show the pupil as a tiny semicircle pendent from the upper lash line; a separate line is added under the brow to suggest the upper lid on the three goddesses but not on the other figures. Save in the case of Eros, the curls over the brow are already being treated as overlapping concentric circles or spirals, but this is not yet being handled completely mechanically, e.g. developing into a double spiral at the front of Athena's brow and in the locks above her ear. The three-quarter view of the nude torso of Aphrodite is competently drawn, although the only muscle detail is the *linea alba* which terminates at the navel. The drapery rendering is, on the whole, both skilful and sensitive. Especially subtle is the treatment of the edge-folds where the thickness of the material is sometimes suggested by not having the lines quite meet. There are, however, shortcomings. Thus the fold-lines fail to be continued through the border of Hera's overfold and there are considerable illogicalities at the back of Aphrodite's drapery beside the right leg of Athena. The left-hand column is not carried through under Paris' right arm and behind his legs. The repetitiveness of the goddesses' pose has a not unpleasing effect. The wings of Eros and the upper Lasa or goddess have their flight feathers shown individually, but only the outer feathers of the coverts. A line inside the leading edge of the coverts suggests the thickness and shape of the wings; this is a solid line on those of the Lasa, but dotted on those of Eros. The garland emerges rather awkwardly from behind the head and wings of Eros.

Fabric and Date. Etruscan, still Early Hellenistic or beginning of Hellenistic. The present three-tier composition, with a child figure on the extension under the main scene and another figure in an upper exergue above it seems to have evolved from slightly earlier northern tanged mirrors of the end of Late Classical and Early Hellenistic date, such as those of the Indiana Group, for which see L. BONFANTE WARREN in *StEtr* XLV, 1977, pp. 149-68, pls. 21-6. On where these were probably made, see E. MANGANI in *BdA* 33-4, 1985, pp. 21-6, figs. 1-7. Mirrors such as GERHARD, *ES*, pl. 257B, PFIFFIG, *Herakles* pl. 23 help bridge the gap between such works and the rather simpler and delicately refined style of the Cambridge mirror. For a suggested much later continuance of such 'raised-border' mirrors, see U. HÖCKMANN in *JdI* CII, 1987, pp. 247-89.

5. Engraved mirror. Figs. 5a-d

Uninventoried. Unprovenanced. Bequeathed in 1891 by S. S. Lewis, who had acquired it in 1884; previously Castellani collection. Carries old stickers, one with old number: '18'. (Fitzwilliam Museum loan register Ant.1991/103.37.)

Cat. Castellani, Paris, 1884, p. 47 no. 421; C. W. KING in *ComCAS* V, 1880-4, p. 192, pl. 5; KLÜGMANN-KÖRTE, *ES*, pp. 90-1, no. 73*a*; T. J. DUNBABIN in *StRobinson* II, p. 1172 n. 57; *Classical Heritage*, p. 15 no. 58; S. J. SCHWARZ in *LIMC* V, p. 232 no. 333, *s.v.*.'Herakles/Hercle'.

Bronze. Crack that apparently developed in ancient hammering and finishing has caused loss of tip of one point. Patina partly green malachite, including some areas of heavy corrosion, partly a greenish black. Has undergone manual cleaning, notably on *rev.* and lower *obv.* of disc, revealing extensive areas of cuprite. Deep centre point of *rev.* in middle of engraved right thigh of Herakles. On technique, see Appendix.

W. 13.15 cm. H. 24.0 cm. H. of disc 14.7 cm. H. of handle 9.4 cm. P.W. at points 3.55 cm. Wt. 230.9 gm. Composition (see p. 54); copper 86.5%, tin 12.32%.

Mirror with a pear-shaped disc but lacking an extension apart from vestigial traces in the form of small projecting points and with a tapered handle cast in one piece with the disc, the terminal in the form of a small, stylized head of a hind (or fawn?).

Obv. Very slightly convex reflecting face bounded by slight channel and narrow moulding of hemispherical section which continue along the neck and grip of the handle; the grip bears a slightly convex flattish raised shape with rounded ends at top and bottom. Edge of disc of projecting oblique channelled section with relief tongue pattern with raised outlines. Lower part of disc engraved with a lean, long-legged dog shown in a half-crouch apparently on rocky ground. For the species of dog, cf. the hounds in the wall-paintings of the François Tomb (GIGLIOLI, *AE*, pl. 264, 1, 3) and, even closer, the greyhound on the Pelops mirror in Paris (REBUFFAT, *Miroir*, no. 47). The engraving is now quite faint, probably due partly to the heavy accumulation of cuprite, more to wear from the repeated ancient polishing of the reflecting face.

Rev. Back of disc very slightly concave, of handle flat. No true extension nor any extension ornament. The border consists of two undulating sprays of ivy with alternate leaves and berries forming a somewhat incomplete garland and beginning at the horizon of the medallion scene. Between the tips of the sprays at the top there is a bunch of berries(?). The groundline is rocky with blades of grass here and there. In the centre stands a curly-haired beardless young man. He seems to be identified as Herakles (Hercules) by the attributes that he has apparently just discarded: a knotty club set obliquely between the rocks behind him, a bow and arrow propped upright in front of him and a lionskin sprawled at his feet and filling the foreground. He is standing at a three-quarter angle beside a winged stallion, gripping the reins on each side with his hands and leaning back slightly, his legs apart, to pull in the horse. The winged stallion is also in three-quarter view, apparently already almost at a halt with his hindlegs firmly on the ground, although his fore-hooves are still raised and his mane and tail still flying. He is fully harnessed and has a thong with two *bullae* around his upper neck and a continuous band of *phalerae* around it somewhat lower down. In the field, between his head and that of Herakles, there is a large eight-pointed rosette-star and between his tail and wings is a dot-rosette of seven tiny circles.

Subject. Herakles (Hercules) training Areion (Ario). The scene can hardly be the capture of the man-eating horses of Diomedes which are sometimes shown winged in Etruscan art (BROMMER, *Denkmälerlisten* I, p. 145), partly because these are more usually depicted as a team and partly because the hero has discarded his weapons and lionskin in readiness to try to mount the horse. He was first recognized as Herakles by KING. The wonder-horse Areion was born of Poseidon and Demeter and received by Herakles as a gift, eventually passing from him to Adrastos who escaped on him when his comrades were slain in the attack by the Seven on Thebes. Current opinion (I. KRAUSKOPF in *LIMC* II, pp. 477-9) seems to be that Areion is always shown wingless, never winged and thus can readily be distinguished from the winged Pegasos in ancient art, a doubtful criterion so far as Greek representations are concerned (cf. T. J. DUNBABIN in *StRobinson* II, pp. 1172-3, 1180-2) and refuted by Praenestine scenes such as this. A second, slightly earlier Praenestine mirror, once also in the Castellani collection, repeats the same scene, with the winged Areion rearing higher and with Herakles wearing his lionskin round his hips as a loin-cloth (GERHARD, *ES*, pl. 334; T. GERICKE in *AM* LXXI, 1951, p. 194). The winged Areion seems also to be shown twice, antithetically, on a Praenestine *cista* in Palestrina (BORDENACHE, *Ciste* I, 1, no. 52), with the accompanying figures of Herakles (a) clubbing an ill-preserved Nemean lion while Hermes carries off one of the lion's victims and (b) newly dismounted (and therefore without attributes) before Zeus and Hera on Olympos for his apotheosis, the two scenes being apparently separated by a silen with straying panthers. For an Etruscan mirror showing Herakles (*Herkle*) riding a misnamed wingless flying horse presumably intended as Areion, see **1** above. The eight-pointed rosette-star may well be more than a filling ornament and indicate a star in the sky, marking the scene as set at night or at dusk or dawn; cf. the similar star on the Eos (Aurora) mirror in Paris (REBUFFAT, *Miroir* no. 41). In that case, the seven-point dot-rosette above the horse's rump might possibly also be astral, signifying that cluster of seven weaker stars, the Pleiades.

Engraving Characteristics. The engraving of the *rev.* is of moderate depth and shows firm, even and skilfully controlled lines, particularly on the figures, but not always quite so smoothly flowing on the garland. The internal detail of the figures is economical, but effective and in an advanced style. Herakles' three-quarter eyes are rendered by two joining lines, one for the upper lash, the other for the pupil, and the slightest of lines suggest the deltoids, *serratus magnus*, iliac crests, groin, knee muscles, etc. A freer use of small lines renders the three-quarter-angle horse's head quite well, although the gap between the eye and the upper lid line gives almost the effect of a human eyebrow. The structure of the shoulder and chest are also well suggested. Coarse hatched shading gives shape to the animal's belly and haunch. The flight feathers of his wings are indicated individually, but the way in which the wings themselves sprout from his shoulders is fudged.

Fabric and Date. Praenestine, Hellenistic. Although some of the characteristics of the horse can be matched in Late Classical work, the proportions and advanced drawing of Herakles would seem definitely Hellenistic. Also advanced would seem the now only vestigial garland border and the debased shape of the mirror itself. The date tentatively suggested here would seem distinctly later than that implied by ADAM (*Miroirs prénestins, passim*) for the small, late, debased Praenestine mirrors. It would also imply that the Bellerophon and Pegasos mirror found in a burial apparently of the 3rd century BC at Palestrina in 1917 (*Roma Medio-Repubblicana*, pp. 276-8 no. 421, pls. 83-9), which seems distinctly earlier than the Cambridge mirror, is actually an heirloom, older than the context in which it was found.

6. Engraved mirror. Figs. 6a-d

Uninventoried. Found at Praeneste in 1872 or slightly earlier. Bequeathed in 1891 by S. S. Lewis who had secured it from Hoffmann at least some months after the Castellani sale in Paris of May, 1884; previously Castellani collection. (Fitzwilliam Museum loan register Ant.1991/103.36.)

W. HELBIG in *Ephemeris Epigraphica* I, 1872, p. 153 no. 168*b*; R. GARRUCCI, ed., *Sylloge Inscriptionum Latinarum Aevi Romanae Reipublicae usque ad C. Iulium Caesarem Plenissima*, Turin, 1877, pp. 162-3 no. 539; H. JORDAN *Kritische Beiträge zur Geschichte der lateinischen Sprache*, Berlin, 1879, pp. 5, 72-8 no. 18; *CII, App.*, no. 926; H. JORDAN in *Hermes* XVI, 1881, p. 233; *Cat. Castellani*, Paris, 1884, p. 50 no. 431; F. BÜCHELER in *RhMus* XLII, 1887, p. 320; H. DESSAU, ed., *CIL* XIV, 1887, no. 4104; E. LATTES, *Le iscrizioni palaeolatine dei fittili e dei bronzi di provenienza etrusca*, Milan, 1892, no. 132; KLÜGMANN–KÖRTE, *ES*, pp. 197-201, pl. 151; R. S. CONWAY, *The Italic Dialects, edited with a Grammar and Glossary* I, Cambridge, 1897, p. 318 no. 297; A. ERNOUT in 'Le parler de Préneste' in *Mémoires de la Société de Linguistique de Paris* XIII, 1905-6, p. 300; H. DESSAU, *Inscriptiones Latinae Selectae* II, 2, Berlin, 1906, p. 952 no. 8564; MATTHIES, *PS* pp. 61, 66, no. A13; T. MOMMSEN, E. LOMATZSCH, eds., *CIL* I^2, 2, 1918, no. 559; E. VETTER, *Handbuch der italischen Dialekte*, Heidelberg, 1953, pp. 337-8 no. 366*m*; E. DIEHL, *Altlateinische Inschriften*, 5th edn, Berlin, 1964, p. 80 no. 770; *Classical Heritage*, p. 15 no. 57, with fig.; DE GRUMMOND, *Guide*, p. 72 n. 31; R. WACHTER, *Altlateinische Inschriften* (*Europäische Hochschulschriften* Ser. XV, XXXVIII), 1987, pp. 117-18 no. 47.

Bronze. Buckled and broken into three fragments; decoration virtually complete, as shown reassembled in figs. 6a, c, but inevitably distorted. Medium-green malachite patina over light layer of cuprite. Light manual cleaning. Modern pale pink infill in *rev.* engraving, but part of mouth of animal and fold of drapery of left-hand figure uncleaned and unfilled; infill probably wrongly added to indicate word punctuation between *r* and *et* near middle of long vertical inscription (apparently natural damage) and probably at end of that inscription; faint vestigial *i* overwritten by *o* of *atos*. Centre point on *rev.*, on buttock of engraved seated woman.

Original W. 17.55 cm. Original H. 32.7 cm. H. of disc and extension 12.7 cm. W. at points 4.15 cm. Wt. 333.9 gm. Composition (see p. 54): copper 88.5%, tin 10.54%.

Pear-shaped disc with concave-sided extension with sharply projecting points; handle cast in one piece with disc and ending in terminal in form of stylized head of hind (or fawn?).

Obv. Disc with originally slightly convex face bounded by slight ridge which also continues along sides of extension and of neck and grip of handle. Edge of disc of projecting oblique profile decorated in relief with summary tongue pattern consisting of row of round-topped grooves. Grip and neck of handle ornamented in relief with elongated floral(?) motif, similar to that of **2**, but lacking scallop-like frill between outer petals at top. Extension apparently originally engraved with palmette flanked by horn-like volutes, now nearly all obliterated by repeated ancient polishing of face.

Rev. Back of disc originally slightly concave, that of handle concave and somewhat irregular. Extension engraved with inverted five-petalled palmette with berries on stalks (or stamens?) either side of central petal. This palmette is flanked by volutes, taking the form of spiral tendrils with berries for eyes and bound together across the stem of the palmette, developing into a line encircling both medallion and exergue. The two undulating sprays of ivy constituting the border decoration also spring from the sides of the volute tendrils and meet at the top under a central squarish rosette. The triangular exergue is occupied by a cluster of acanthus growing upwards from the stem of the extension palmette and with a large trumpet-shaped flower in the middle, flanked by tendril-like stalks ending in clusters of berries(?). To judge from its rocky outcrop and tuft of grass, the medallion scene is set out of doors. Behind the figures to the right of the composition there is a rectangular construction shown in three-quarter view. On the evidence of its massive, projecting plinth and crowning, both articulated by rounded mouldings, this would seem to be either a statue-base or an altar and very probably the latter, since the animal shown sprawled on top of it certainly does not seem a statue. This animal has hitherto and probably rightly been interpreted as a dog, presumably of the long-haired Maltese breed, but its awkward, fussy engraving does not entirely preclude other possibilities, in particular that of a small wild boar. In front and obscuring one end of both animal and altar there is the slender figure of a girl, dressed in a long garment which is apparently not open at the side and thus presumably a form of sleeveless *chiton*, or tunic, with large disc-shaped fastenings at the shoulders and worn with a long girdled overfold. She also wears shoes and an earring with a plain round pendant and, with her right hand, she holds up a mirror, apparently one with a circular disc and not of the pear-shaped form of the present example. Beside her is the inscription, *acila* (presumably 'maidservant'; see below). The centre of the composition is occupied by a pair of lovers embracing and gazing at each other. A beardless curly-haired young man is seated on an outcrop of rock, with a *himation* wrapped around his legs and hips. On his lap is seated a young woman, naked apart from her shoes and a wristlet and double or triple armlet on her left arm. Adjacent inscriptions (see below) name the man as *Mexio* and the woman as *Fasia*. To the left of this couple is a further woman with her head turned back to gaze at the lovers and the girl with the mirror. Otherwise naked, she has a short, elaborately bordered cloak with a weight at one corner draped across her shoulders and hanging down her back, with one edge clasped in her right hand, and also wears shoes, bracelets, a choker necklet with a crescent-shaped *bulla*(?) and an earring with three pendants probably of Etruscan type (cf. HIGGINS, *Jewellery*[2] pl. 42*c*). See below on the inscription beside her.

Inscriptions. All written from left to right. Almost vertical behind left shoulder of girl with mirror:

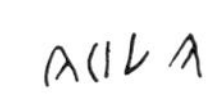

acila

Horizontal above head of seated young woman:

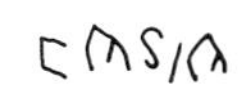

fasia

Almost vertical behind head of seated young man:

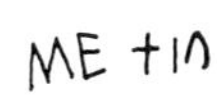

mexio

Vertical near edge of medallion beside left-hand woman:

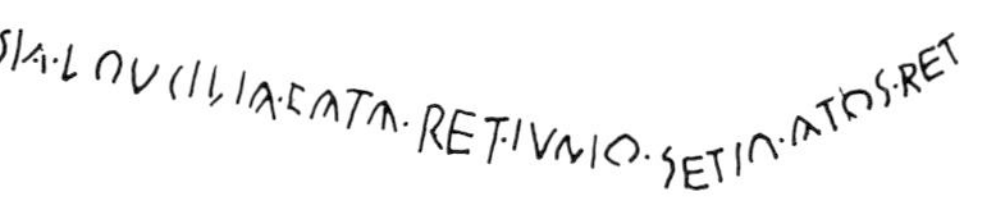

ceisia · loucilia · fata · ret . iunio · setio · at⟦i⟧os ·ret

In dealing with these difficult inscriptions the author is deeply indebted to generous help from Dr R. Wachter and also wishes to acknowledge advice from Professor R. G. Coleman and Miss J. M. Reynolds. *Acila* is apparently the classical Latin *ancilla*, 'maidservant'; for the omission of the *n*, see WACHTER, *Altlateinische Inschriften* under nos. 63*c*, 122*a*, etc., and in index, p. 548. *Fasia* and *Mexio* appear to be personal names, but without obvious classical counterparts, although, in correspondence with the author, Dr Wachter has suggested that *Mexio* might be *Mexio(s)* and be an *-isios* derivation from *Maicios* (classical *Maecius*) with monophthongization of *ai* to *e* and syncope of the middle vowel; *Fasia* might conceivably be linked with the ethnic *Phasias* ('woman of Phasis', used, e.g. of Medea). In the past a number of scholars (e.g. KÖRTE, DIEHL, CONWAY, ERNOUT, DESSAU and VETTER) have proposed reading the *x* of *Mexio* as *t* and the name as the equivalent of the classical *Mettius* and some (e.g. CONWAY, ERNOUT and VETTER) have also taken the *f* of *Fasia* as *c* and her name as *Cassia*. The *f* recurs in the longer inscription in *fata* (read *cata* by VETTER), but as these inscriptions otherwise consistently use normal forms of *c* and *t* it seems much more likely that these other letters have their more natural force of *f* and *x*, as reinstated by WACHTER. Relatively few comprehensive interpretations have been offered for the longer inscription. KÖRTE suggested that it might have funerary votive force, signalling a return by relatives to the dead of a mirror she had used in her lifetime, and reading *Ceisia Loucilia Fata(i) ret(tulit). Iunio(s) Setio(s) ato(i)s ret(tulit). Ato(i)s* was regarded as linked to the classical *atavis*. The inscription, however, seems to be in the same hand as those naming the figures in the medallion scene and is probably an integral part of the medallion decoration and not a subsequent votive addition. VETTER thought that *Ceisia Loucilia* might be the recipient of the mirror and *Iunio(s) Setio(s)* its donor. Most recently WACHTER has raised the possibility of the inscription's being a more oblique comment, reading *Ceisia Loucilia fata re T. Iunio Setio atoiret*. This involves emendations to the text and its word divisions. In correspondence with the author, WACHTER has indicated that no such emendations were intended and that the reading was based on a faulty photograph and is not borne out by better pictures now in his possession and recent autopsy in Cambridge. It seems best now to try to analyse the text as it stands, again with WACHTER's help. *Ceisia Loucilia*: probably *praenomen* (cf. WACHTER, *op.cit.*, p.117 n.286) and *nomen*. *fata*: if with its classical force, probably 'oracles' or 'fates' or, but less meaningful in this context, the past participle of *for*. *ret*: perhaps not a shortening, e.g. of *ret(tulit)*, as taken by KÖRTE, since such a verb abbreviation is almost unknown in early Latin; possibly a verb such as *reddit* with syncope or, more probably, an active form of *reor*, lost from classical Latin apart from its passive past participle, *ratus*, from which one may perhaps infer its meaning as 'calculate', 'evaluate'. *Iunio Setio*: if *setio* is also a proper name, either *nomen* and *cognomen*, with the *praenomen* omitted, or an unique instance of *Iunio(s)* as a *praenomen* with the *nomen*, *Se(n)tio(s)* (for the omission of the *n*, cf. *acila* above); VETTER suggested that *Setio(s)* might, as a *cognomen*, be based on an ethnic of the town of Setia, but one might in that case expect something closer to the classical *Setinus*; presumably either nominatives with the *-s* omitted, as possibly also in *Mexio(s)*, or oblique cases, probably datives, with the final *-i* omitted. *At*⟦*i*⟧*os*: perhaps an unknown common noun lost from classical Latin or the Sabine proper name, *Attus*; the *i* is an erasure, thus precluding the Latin name, *Atios*. The first half of the inscription thus seems to read:'Ceisia Loucilia evaluates the oracles' or 'their fates' (i.e. presumably those of Mexios and Fasia); it may thus allude to the subject-matter of the medallion and identify the adjoining and otherwise unnamed left-hand figure as *Ceisia Loucilia*. The second half of the inscription, however, names one or more people not ostensibly shown in the medallion scene. If *atos* be a proper name, it may possibly read:'Atos evaluates them (i.e. the oracles or the fates of Mexios and Fasia) for Iunios Se(n)tios'. The identity of these two personages must remain a matter for conjecture (as noted above, VETTER suggested that the inscription may have implied that the mirror was intended as a gift from *Iunios Setios* to *Ceisia Loucilia*)

Subject. Embracing lovers, named as *Mexio(s)* and *Fasia*, in the presence of a maidservant with a mirror and another woman. As long as there is no consensus about the significance of its main inscription the interpretation of this scene must remain somewhat problematical. The author is accordingly very grateful for the advice of various colleagues, in particular Dr J. Swaddling, Dr S. Haynes, Dr G. Bordenache Battaglia and Dr H. Salskov Roberts. With the exception of GARRUCCI, who considered the lovers to be Prometheus and Asia, most scholars have regarded this scene as showing contemporary Praenestine life. Although it clearly takes place out of doors, most have thought the setting to be domestic, although VETTER considered *Fasia* to be a *hetaira*. The long *Ceisia Loucilia* inscription has tended to be regarded as divorced from the left-hand figure beside it who is otherwise unnamed and this figure has been interpreted as Aphrodite (Venus), e.g. by KÖRTE and VETTER. But their hairstyle, with pronounced 'ponytails', may mark all the women as contemporary mortals, although, on the evidence of the Villa Giulia mirror cited below, this feature is possibly rather to be viewed as a characteristic of a brief phase of the artist's work. The main accompanying inscription may allude to oracular responses, but the present scene does not seem to be obviously linked to the famous oracle of Fortuna at Praeneste, possibly shown on an engraved *cista* in Rome (*Roma Medio-Repubblicana*, no. 422, pls. 79-80; to the contrary, H. RIEMANN in *RM* XCI, 1984, pp. 389-98). Perhaps some other form of prophecy may be involved. The maidservant with a mirror has usually been considered a stock figure of domestic and toilet scenes, but what if the mirror is being used in some form of cathreptomancy, κατοπτρομαντεία (cf. L. BALENSEIFEN, *Die Bedentung des Spiegelbildes als ikonographisches Motiv in der antiken Kunst*, Tübingen, 1990, pp. 174-209), possibly to determine the future of the lovers, *Mexio(s)* and *Fasia*? In that case the left-hand figure might indeed be *Ceisia Loucilia* herself fulfilling her role. The plinth and sprawled animal have usually been regarded as minor incidental details, but in this setting they might be highly significant as altar and, possibly, victim. If the animal is correctly interpreted as a dog, then the deity invoked should be Hekate or some Italian counterpart. But at this stage these suggestions must remain tentative. It is to be noted that the other closely related mirrors cited below are uninscribed and are decorated with mythological scenes or conversation groups, possibly strengthening the view that the present engraving may have been specially commissioned.

Engraving Characteristics. The engraved lines are narrow, but firm and sure, particularly in the main outlines. The strength of the engraving lies in the shorter lines of equal weight indicating the internal detail, which have a remarkable aptness and economy in suggesting both the play of drapery folds and the bone structure and musculature of the bodies. The artist shows strong individuality, particularly in his somewhat old-fashioned profile heads with their characteristic noses, mouths and chins. The eyes, full-face on the three-quarter head of *Mexio(s)* but otherwise in profile, have the pupils indicated by a short vertical stroke or an U-shaped line and often have the upper-lid line set rather high, almost like a second eyebrow. The perspective treatment of the altar or plinth is not entirely successful, but that of the maidservant's mirror is more competent. The flower and acanthus leaves of the exergue are shown in perspective with hatched shading.

Fabric and Date. Praenestine, end of Late Classical. Although dated much earlier by MATTHIES, this mirror must come down into the Late Classical period of Praenestine engraving, on the evidence of its advanced exergue floral ornament. Indeed, it may fall quite late in that phase, although, with the exception of the maidservant, the figures still show classical proportions. The wavy hair of all save the left-hand figure is rendered as a pattern of groups of concentric arcs following the same convention as that used for the smaller feathers of Etruscan rf. duck-*askoi* of the late 4th and beginning of the 3rd centuries BC (M. HARARI, *Il'Gruppo Clusium' nella ceramografia etrusca*, Rome, 1980, pp. 45-62, 131-56), both those of the Clusium Group (*ibid.*, pls. 24-32, 35) and of other fabrics (*ibid.*, pls. 33-6, 38-9), possibly including Faliscan; this convention is adapted for human hair above the brow by the Volterran Hesione Painter (BEAZLEY, *EVP* pp. 122, 124-7, pl. 29, 6), possibly mirroring the emergence of the stylizations characteristic of Hellenistic Etruscan engraving (e.g. on **4**, **10**, **17**, **18**, **19**). We may call the engraver the Lewis Master; also by him at the same stage in his career is: Rome, Villa Giulia inv. 24081, (GERHARD, *ES*, pl. 210; R. D. DE PUMA in *RM* LXXXVII, 1980, p. 23, fig. 8). A number of other mirrors showing virtually the same border and floral decoration would seem to be products of the same workshop: (1) Dresden, Antikensammlung inv. ZV32 (KLÜGMANN-KÖRTE, *ES*, pl. 46; *CSE* DDR. 2, no. 12) (2) London market (*Sotheby Catalogue*, 17-18 July 1985, no. 269); (3) Madrid, Museo Arqueologico inv. 9824 (J. M. BLASQUEZ in *AEArq* XXXIII, 1960, pp. 148-52, figs. 2, 4). A number of these mirrors have already been linked in the past (MATTHIES, *PS* pp. 61, 66-7; ADAM, *Miroirs prénestins* p. 84, fig. 8). Their style is in various ways near to that of the Lewis Master, but the relationship is not clear. Some other Praenestine mirrors with totally different border and floral ornament show possible links with the Lewis Master's work, e.g. the Dionysiac scene on one in the Musée de Mariemont (*CSE* Belgium, 1 no. 24) which reveals a similar hair rendering by a pattern of groups of concentric arcs and a possibly related treatment of one profile head, but which seems otherwise distinctly different in style.

II The Fitzwilliam Museum

The Fitzwilliam Museum holds the main art collections of Cambridge University. It was endowed in 1816 by the bequest of its founder, Richard, seventh Viscount Fitzwilliam of Merrion, and its present main structure, the Founder's Building, the lower floor of which still houses the antiquities displays, was opened in 1848. The first Etruscan mirrors to be acquired, **8** and **9**, were received in 1864 when the University bought the entire collection of Greek and Roman coins and antiquities formed by the great topographer of ancient Greece, William Martin Leake (1777-1860). The bequests of F. McClean in 1904 and of the two artists, C. S. Ricketts and C. H. Shannon, in 1937 each added one more item, *viz.* **10** and **7**, and two more (**11** and **12**) have recently been secured by purchase, one with help from the National Art-Collections Fund and the other by the Friends of the Fitzwilliam Museum. New acquisitions at the Fitzwilliam are described in the *AnRepFitzwilliam* whose text also appears in the *Reporter* but which is cited here in the form of the separate illustrated edition circulated to subscribers. Unfortunately, up to the 1950s, the larger benefactions tended not to be described item by item. Normal Fitzwilliam Museum inventory numbers are made up of three components, of which the first is the departmental prefix and the last the year of acquisition; the middle component is the serial number of the item in that year and under that department. Two departmental inventories are involved, those of Egyptian and West Asiatic Antiquities (prefix 'E') and of Greek and Roman Antiquities (prefix 'GR'). This account omits accessions made since its last complete revision in 1986.

7. Relief handle probably from mirror. Figs. 7a-e

Inv. E.101.1937. Unprovenanced. From the collection formed jointly by C. S. Ricketts and C. H. Shannon and bequeathed by them in 1937; although apparently Etruscan, received with the Ricketts and Shannon Egyptian collection (whence the anomalous inventory prefix).

Unpublished.

Bone. Hollow cylinder forming grip of handle; other parts missing. Cracked in half vertically (reglued in modern times); lower end rotted away; otherwise in good condition with faint traces of red pigment surviving in isolated recesses at edges of several of lotus blossoms and in mouth of one of Gorgoneia; green stains from bronze of mirror(?) once fitted in handle.

D. between 3.8 and 4.0 cm (not perfectly circular). P.H. 8.9 cm. Original H. *c.* 9.25 cm. Thickness of wall 0.4-0.5 cm. Wt. 83 gm.

Hollow cylinder with interior knife-hollowed and exterior carved freehand with the following relief decoration. At top (and probably also originally at lost bottom) plain projecting border. Whole of remainder of surface occupied by continuous pattern of four Gorgoneia and eight lotus blossoms, the latter arranged in pairs to frame the Gorgon heads. The Gorgoneia have flat faces, relief outlining around the eyes, broad noses and large mouths with exposed teeth and protruding tongues; their long, fairly straight hair is parted in the middle.

Subject. Heads of Gorgon, Medusa (Metus), framed by lotus blossoms. For Metus and the Etruscan use of the Gorgoneion, see PFIFFIG, *Religio*, pp. 317-19.

Fabric and Date. Etruscan, Early Classical. The total effect of the decoration seems not far removed from Late Archaic, but the centrally parted stranded hair of the Gorgoneia seems more likely at the start of the Early Classical Period. For the style of the Gorgoneia, cf. those of the paintings of the Tomba della Pulcella in Tarquinia (GIGLIOLI, *AE*, pl. 208, 1) and of an engraved mirror in Paris (REBUFFAT, *Miroir*, no. 63). For such bone mirror handles with relief decoration, see Y. HULS, *Ivoires d'Étrurie*, 1957, pp. 83-7, 91-2, 110-16, 203-8, pls. 49-53; S. S. WEINBERG in *Muse* IX, 1975, pp. 25-33; U. LIEPMANN, *CSE* W. Germany 2, nos. 20, 22-3.

8. Engraved mirror. Figs. 8a-d

Inv. GR.12.1864. From Vulci (its provenance, given as firm in the earliest mentions of it by GERHARD and ROULEZ and in *CII*, seems subsequently to have been overlooked). Purchased by Cambridge University, 1864; formerly Gerhard and Leake collections.

E. GERHARD in *AbhBerlin*, 1836 p. 348 n. 180 (= *Gesammelte akademische Abhandlungen und kleine Schriften* I, 1866, p. 131); J. ROULEZ in *AnnInst* XV, 1843, pp. 215-17, pl. F: G. MINERVINI in *BullArchNap* Ser. I, III, 1845, p. 48; J. OVERBECK, *Gallerie heroischer Bildwerke: die Bildwerke zum thebischen und troischen Sagenkreis*, Brunswick, 1853, pp. 84-8, pl. 3, 3; GERHARD, *ES* II, pp. 171-2, pl. 178; *CII*, no. 2152, glossary pp. 99, 206, 1874; L. G. ELDRIDGE in *AJA* XXI, 1917, p. 385 n. 2; C. ROBERT, *Archäologische Hermeneutik*, Berlin, 1919, pp. 218-19 fig. 173; J. SCHMIDT in ROSCHER, V, p. 1401 *s.v.* 'Tydeus'; FIESEL, *Namen*, pp. 27 n. 112, 53 n. 202, 97; G. A. MANSUELLI in *StEtr* XX, 1948-9, p. 85; G. B. MONTANARI in *EAA* I, p. 72 *s.v.* 'Adrasto', fig. 110; DE SIMONE, *Entleh.* I, pp. 16 no. 6, 29 no. 2, 122 no. 6; G. RONZITTI-ORSOLINI, *Il mito dei sette a Tebe nelle urne volterrane*, Florence, 1971, pp. 11-13 with fig.; I. KRAUSKOPF, *Der thebanischen Sagenkreis und andere griechischen Sagen in der etruskischer Kunst*, Mainz, 1974, pp. 46, 108 no. Sie 5, pl. 24; M. DAUMAS in *REA* LXXVII, 1975, p. 24, pl. 4; BROMMER, *Denkmälerlisten* III, p. 6 no. 1, *s.v.* 'Adrastos'; FISCHER-GRAF, *Vulci*, pp. 47 n. 485, 70, 72 n. 702, 95-7 no. V69, 110-11 fig. 3, 112, 115, pl. 26, 2; I. KRAUSKOPF in *LIMC* I, p. 697 no. 28 *s.v.* 'Amphiaraos', pl. 559; J. P. SMALL, *Studies Related to the Theban Cycle on Late Etruscan Urns*, Rome, 1981, p. 145.

Bronze. Areas of bright and pale green malachite(?), overlaid by brownish incrustation, survive over parts of tang and edges and much of *obv.*; extension area of *obv.* cleaned manually to cuprite layer or beyond to bare metal; whole of *rev.* of disc and extension heavily cleaned and repatinated a brownish black; *rev.* much pitted by corrosion, obliterating engraving at a few points. Large, deep centre point on *rev.* on left upper thigh of engraved figure of Tydeus. White-ink infill in *rev.* engraving, showing various minor inaccuracies; published nineteenth-century facsimile drawings also reveal errors and unjustified restorations. On technique see Appendix.

D. 18.0 cm. H. 25.7 cm. H. of disc and extension 21.0 cm. W. of bottom of extension 3.5 cm. H. of tang 4.7 cm. Wt. 472.2 gm. Composition (see p. 54): copper 87.5%, tin 10.59%.

Circular disc with concave-sided extension and tapered tang.

Obv. Slightly convex reflecting surface bounded by plain moulding of rounded section continuing along sides of extension. Beyond this the edge, of channelled slightly offset oblique profile, carries relief decoration of a tongue pattern with raised outlines. Extension engraved with pair of double volutes flanking a five-petalled palmette with small circles either side of its tip and with half-palmettes against the upper volutes. Originally there was a further crowning palmette above, flanked by double-volute tendrils, but much of the upper part of this decoration has been obliterated in the repeated ancient polishing of the mirror face and what survives is now very faint, although the lower part of the extension engraving remains quite deeply cut.

Rev. On the extension there is an inverted lotus blossom with laterally striped(?) petals, set under a pair of elongated double volutes flanking a nine-petalled palmette with berries on stalks (or stamens?) either side of the central petal. To each side of this there emerges an undulating spray of ivy with alternating leaves and berries which encircles half the medallion as a border. Where they meet at the top these are overlaid by a bow-shaped element with a cross-hatched interior and reserved border, perhaps intended as tying them together into a garland and if so constituting a rare variant for the Herakles' knot met with in this position on a number of late Vulci mirrors. As noted by ELDRIDGE, the same butterfly-bow(?) motif recurs on another late Vulci mirror now in Boston: COMSTOCK–VERMEULE, *Boston Bronzes,* no. 377; FISCHER-GRAF, *Vulci* no. V91.

The medallion scene is separated from the floral ornament above the extension by a broad band filled with alternately cross-hatched and reserved triangles, their arrangement perhaps meant to echo the bow shape at the top. The upper edge of the band forms a base-line for the scene. This shows three men, the middle one standing, while his companions are seated on outcrops of rock rendered obliquely in a kind of high-eye-level perspective. The man on the left is named in the accompanying inscription (see below) as *Atrste* (Adrastos). Bearded and with generous locks, he sits with his gaze directed at his companions, his raised left hand gripping his vertical spear and his right arm near his side, the half-open hand perhaps intended to be supported against the rock. His left foot rests on the rock and his right on the ground below it, extending into the decorative zone below. He wears a short under-*chiton* fastened along the upper arms to form sleeves and fitted with a drapery weight (both details most clearly visible below his right shoulder). Over this he wears a leather(?) cuirass with two rows of *pteryges*, possibly intended as lined throughout, to judge from the banding of the edges. A short *chlamys* is fastened around his neck and hangs down his back, with a drapery weight at the corner visible under his left arm. His feet appear bare, with an awkward oblique modelling line on the right foot. The central figure is that of a young man shown standing facing him with his body at a three-quarter angle and his weight more particularly on his slightly advanced right foot. He is named as *Tute* (i.e. Tydeus, Adrastos' son-in-law) in the accompanying inscription and is naked save for a *chlamys* with banded edges pinned around his neck and hanging down his back, with drapery weights at its four corners. His hair extends down to the nape of his neck and he wears sideburns but is apparently otherwise clean-shaven. In his lowered left hand he carries a spear, held vertically with its butt clear of the ground and gripped between his first and second fingers and thumb. He is apparently in conversation with Adrastos and also looking at an armlet with three pendent *bullae* which he is holding looped around the palm of his right hand. The right-hand figure faces the other two. He is named in the accompanying inscription as *Amphiare* (i.e. the seer, Amphiaraos). He is clean-shaven and his hair hangs down on to his shoulders, bound by a narrow head-band. Two fingers of his right hand are raised to his part-open mouth and his left hand rests on the hilt of his sheathed sword resting vertically on the ground beside him, the exterior of its scabbard apparently bound with overlapping bands of leather and its unfastened carrying-thong seemingly trailing loosely to the left while its suspension loop projects to the right. He wears a *himation* wrapped around his hips and legs and with one corner, probably bearing a drapery weight, carried around his back and hanging down over his left shoulder, while another corner, also weighted, is brought across his lap and trails down behind his right leg. On his feet he wears substantial shoes with a band or lining overlap along their upper edge. On his left upper arm he wears an armlet with three pendent *bullae* of the same kind as that in Tydeus' right hand.

Inscriptions. All written from right to left. In a vertical, roughly rectangular frame, or cartouche, beside left figure's left elbow and spear:

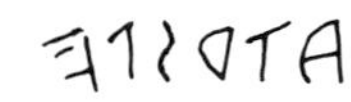

Atrste

In an almost horizontal quadrilateral cartouche to the left of the brow of the central figure:

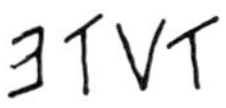

Tute

In an almost horizontal roughly rectangular cartouche, above the head of the figure to the right:

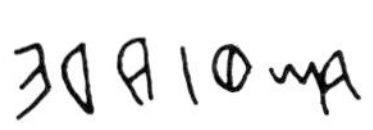

Amϕiare

A and *m* are in ligature. The use of cartouches, as if the inscriptions had been added on *pinakes*, or tablets, is a characteristic of the mature Vulci style as identified by FISCHER-GRAF (*Vulci*, p. 72 and nos. V42, V45, V46, V47, V59, V62, V70, V76, V94, pls. 18, 19, 21, 26), but is far from invariable there and is met with on other fabrics as well.

Subject. Adrastos (*Atrste*), Tydeus (*Tute*) and Amphiaraos (*Amphiare*). The *bulla*-armlet held by Tydeus is clearly crucial to the interpretation of the scene. It was formerly identified (e.g. by ROULEZ) as the necklace of Harmonia used to bribe Eriphyle, the sister of Adrastos and wife of Amphiaraos, but was re-interpreted as an *armilla* by OVERBECK and GERHARD and its gift viewed by the latter as a token of alliance. Some scholars (e.g. RONZITTI-ORSOLINI, DAUMAS) have regarded the figures as excerpts from a larger composition showing all the Seven and, linking the mirror with Etruscan scarabs with analogous but not identical scenes (ZAZOFF, *Skarabäen* pp. 50-52), derived them all from a Greek painting by Onasias at Plataea known from Pausanias (IX.4.2), although the importance of the *bulla*-armlet in the mirror group would seem to gainsay a Greek prototype. ROBERT viewed the mirror composition as an Etruscan misinterpretation of a quite different scene from Greek mythology. Others have regarded it as a mere conversation scene (KRAUSKOPF) or a re-adaptation of an established iconographic grouping (FISCHER-GRAF). Clearly such *armillae* as shown here were highly valued items of jewellery. BEAZLEY has also revived the important suggestion that when heroes donned *bulla*-armlets given in this way it may have been in preparation for some hazardous undertaking (*EVP*, pp. 33-5, in connection with the Argonaut Group bell-krater in Florence; less attractive seems the interpretation of A. RUMPF in *JHS* LXXI, 1951, pp. 168-71). The scene may thus perhaps show Tydeus agreeing to undertake his mission to Thebes on behalf of the Seven, a journey that was to result in a gory ambush, as possibly here foreseen by Amphiaraos, and as told by Homer in *Iliad* IV.

Engraving Characteristics. The depth of the engraved lines on the *rev*. is at best now only moderate and often very shallow indeed, having presumably been affected by surface loss in both ancient and, more especially, modern cleaning, with the detail completely destroyed at some points, including parts of the heads of Tydeus and Amphiaraos. The main lines of the bodies are firm, and the eyes are shown in true profile with a rather large oval to the front (apparently representing the iris rather than the pupil) as preserved for Adrastos and Amphiaraos who also have some separate indication of the upper lid. The drapery is less competently handled than the bodies, with the *chlamydes* around the necks of Adrastos and Tydeus confused and fussy and the *himation* folds around Amphiaraos' calves unnaturally stylized. Cross-hatching and various kinds of hatched shading are employed. Coarse hatched shading occurs on the lower parts of the rocks and fine body hatching on the flesh of the figures and on Adrastos' cuirass and, at one point, on his *chiton*. Musculature is indicated by a combination of this and solid engraved lines, giving almost the effect of two densities of shadow, to be seen to best advantage on the pectorals, *rectus abdominalis* and leg muscles of Tydeus. Small stamped circles are used for the nipples of Tydeus and Amphiaraos and a row of dots appears on the *bulla*-armlets and parts of the hems of Adrastos' *chlamys* and Amphiaraos' *himation* (this last feature may have been more widely employed, but cannot be confidently distinguished elsewhere because of the heavy pitting).

Fabric and Date. Etruscan, made at Vulci, Late Classical. Assigned by FISCHER-GRAF to her Late Classical Group of Vulci mirrors (for which see also **25** and cf. **11**), which, in her case, implied a date no earlier than the last third of the 4th century BC. Despite residual classical echoes (e.g. in the head of Adrastos) this would seem to be fully justified by the developed style. The mirror's provenance from Vulci seems also significant.

9. Engraved mirror. Figs. 9a-d

Inv. GR.12*a*.1864. From Caere (Cerveteri). Purchased by Cambridge University, 1864. Formerly Carosi collection, Rome, and Leake collection. The documentation of this mirror rests in part on its identity with GERHARD, *ES*, pl. 32, 2, which appears to be the same in all respects save that GERHARD's drawing omits the Lasa's left breast.

GERHARD, *ES*, III p. 11, pl. 32, 2; G. A. MANSUELLI in *StEtr* XIX, 1945-7, p. 62, 'Maestri delle Lase e dei Dioscuri' no. 7; E. MANGANI in *BdA* 33-4, 1985, p. 35 no. 6.

Bronze. Handle and lower part of extension broken away. Patina mostly thin (apart from small localized patches with considerable underlying cuprite), partly green malachite(?), partly brownish black, but with large areas of the surface virtually free of corrosion and apparently protected by a tin oxide patina. Centre point on *rev*. beside *linea alba* of engraved Lasa. Modern white infill in *rev*. engraving. On technique, see Appendix.

D. 11.6 cm. P.H. 14.6 cm. P.W. at extension 2.6 cm. Wt. 116.5 gm. Composition (see p. 54): copper 86.0%, tin 12.68%.

Circular disc with flaring concave-sided extension, its missing handle originally cast in one piece with the disc.

Obv. Small, highly convex reflecting face bounded by channel and beaded ridge, both of which continue along sides of extension. Plain edge of offset oblique profile. What survives of extension preserves upper part of engraved floral ornament consisting of central petal with centre-rib and dotted surface and, to either side, dots perhaps originally flanking edgeways-on(?) side petals, the tip of one of which is preserved to the right (for complete motif, cf. **22** and *CSE* Bologna I, no. 33). The engraving remains deeply cut and the beading crisp.

Rev. Concave back engraved with winged woman advancing to the spectator's left in the so-called *Kreuzschritt*, her left arm akimbo with the back of the hand apparently resting behind her left hip and her right hand holding a slender curved object between thumb and forefinger. She is naked apart from a Phrygian cap and shoes, perhaps intended as tied on with thongs passing under the instep.

Subject. Lasa. For such figures on mirrors of this class, see RALLO, *Lasa*, *passim*. Where they carry attributes, these are usually an *alabastron* and a toilet stick (for the latter, formerly misinterpreted as hairpins, see p. 00), but a small number hold the sort of curved object shown here: (1) Florence, Museo Archeologico (H. SALSKOV ROBERTS in *ARID* XII, 1983, p. 39 fig. 17*c*); (2) Volterra, Museo Guarnacci (*ibid*., fig. 17*d*); (3-4) Viterbo, Museo Civico (EMILIOZZI, *Collezione Danielli*, nos. 592, 593); (5) Nijmegen,

Rijksmuseum Kam (*CSE* Netherlands, no. 25); (6) 'Meer' collection (*ibid.*, no. 31); (7-8) Brussels, Musées Royaux (LAMBRECHTS, *Mir. Mus. Royaux*, nos. 48, 54); (9) Copenhagen, Danish National Museum (*CSE* Denmark 1, no. 7); (10) London, British Museum GR 1916.6-1.30, unpublished; (11) London market (*Sotheby Catalogue*, 9 July 1984, no. 492); (12) New York, Metropolitan Museum (RICHTER, *MMBronzes*, no. 822); (13) Copenhagen, Thorwaldsen Museum (T. MELANDER in *Bronzes hellénistiques et romains*, p. 164, pl. 99, 10); (14) Stuttgart, private collection (*CSE* W. Germany 3, no. 25); (15-16) Paris, Louvre (*CSE* France I, 1, nos. 20, 21); (17) I. WEHGARTNER in E. SIMON, *Die Sammlung Kiseleff im Martin-von-Wagner-Museum*, Würzburg, 1989, pp. 217-18, no. 362, pl. 127. They have usually been interpreted as bell-shaped flowers cupped in the hand, since such Lasae are occasionally shown holding flowers (e.g. D. REBUFFAT-EMMANUEL in *Mélanges de philosophie, de littérature et d'histoire offerts à Pierre Boyancé*, Paris, 1974, pp. 621-4, pl. 2); other suggestions have been hairpins, i.e. toilet sticks (EMILIOZZI), or sponges or wads (LAMBRECHTS). The Cambridge mirror makes the important contribution of indicating the relatively slight thickness of the object. Possibly some common item of adornment is being shown at an unfamiliar angle, perhaps a frontlet for the hair seen from its top or bottom edge. The Ann Arbor Lasa (*CSE* USA. 1, no. 2) holds a much smaller round object (a ring?).

Engraving Characteristics. The cutting is firm and moderately deep, but very summary. A slightly differing preliminary line appears beside the small of the Lasa's back. Her eye is reduced to a simple leaf-like outline, her breasts are shown in profile and her *linea alba* is indicated, terminating at two small flesh folds at about navel level.

Fabric and Date. Etruscan, Hellenistic. On the basis of associated tomb-groups and allowing that some examples remained in use much later, REBUFFAT-EMMANUEL has proposed the 3rd century BC as the main date range for the production of mirrors of this class (REBUFFAT, *Miroir*, pp. 577-91) and SASSATELLI (*CSE* Bologna I, pp. 49-50) has added further such evidence mostly confirming these conclusions although slightly raising the date of the earliest examples, while MANGANI (*loc. cit.*) has tried to define where they were made; some would seem, however, to have continued in use in the 2nd century (H. SALSKOV ROBERTS in *ARID* XII, 1983, pp. 31-53; U. HÖCKMANN in *JdI* CII, 1987, pp. 247-89, especially pp. 260-1). In listing many of the examples, MANSUELLI (*StEtr* XIX, 1946-7, pp.47-8, 62-5) does not attempt to distinguish between the many humble artists that decorated them. The engraver of the present mirror may be named the Leake Lasa Master; of the examples cited above as showing the same attribute, nos. 4 and 5, in Viterbo and Nijmegen, would seem also to be by him and one of the Paris mirrors, no. 15, to be near him.

10. Engraved mirror. Figs. 10a-d

Inv. GR.19.1904. Unprovenanced. Bequeathed by F. McClean in 1904.

O. M. DALTON, *Fitzwilliam Museum, McClean Bequest: Catalogue of the Mediaeval Ivories, Enamels, Jewellery, Gems and Miscellaneous Objects bequeathed by Frank McClean, M.A., F.R.S.*, Cambridge, 1912, p. 120 no. 92.

Bronze. Upper part of disc slightly cracked, with small fragment missing. Retains original deep green malachite patina with small isolated patches of light green and small surface eruptions and accretions; only limited manual cleaning, with parts of *obv.* and *rev.* of disc taken down to underlying cuprite layer; also some small patches of apparently original bright metal on *obv.*, probably protected by tin oxide patina. Faint vestigial centre point on *rev.* above crotch of largely nude engraved woman. Modern French-chalk infill in rev. engraving. On technique, see Appendix.

D. 12.25 cm. H. 26.2 cm. H. of disc and extension 15.6 cm. H. of handle 10.6 cm. W. at extension points 3.7 cm. Wt. 345.5 gm. Composition (see p. 54): copper 85.5%, tin 12.99%, lead 1.107%.

Circular disc with concave-sided extension with sharply projecting points and with handle cast in one piece with disc and decorated all round in relief. Handle ends in terminal in form of stylized head of ram with horns encircling ears; middle part of grip occupied by astragal-like motif with obscure motifs below and open flower above, its crinkly-edged petals framing neck of handle and bottom of extension.

Obv. Small highly convex reflecting face bounded by small half-round moulding between narrow channels and, outside this, by radiating pattern of rounded ribs in relief extending around edge, which is of bulbous raised profile; all these features also continue along sides of extension, whose inner part carries engraved decoration in form of flower with open outer petals tucked inside three-dimensional ones of handle, but with its tall upper petals more schematically rendered as well as forked stamen(?) and crowning leaf-like pistil(?). Lower part of this engraving still deeply cut, upper part now much fainter.

Rev. Concave medallion area of disc is surrounded by raised flat border zone (whose plane continues slightly higher through extension) and, raised again above this, by slightly angled outer band which is prolonged along sides of extension. Extension engraved with stylized elongated flower (or plant), its outer petals (or leaves?) having volutes at bottom, pronounced centre-ribs and oblique side-ribs. Border zone engraved with garland of regularly arranged, close-packed, pointed laurel(?) leaves, probably to be understood as attached to an underlying backing, and encircled at four points by bands of a kind which BEAZLEY has shown to be derived from the *bakchos*-rings used to fasten leafy sprays at the Eleusinian Mysteries (*NumChron* Ser. VI, I, 1941, pp. 1-7). Here, however, the distinctive Union-

Jack or St Andrew's-cross ornament of the *bakchos*-rings has been replaced by cross-hatching within reserved borders of a kind more analogous to, e.g., the 'butterfly bow' of **8**. Within this garland border the medallion is engraved with a four-figure scene apparently set out of doors in front of a building, part of whose architrave and cornice seem to be indicated above the heads of the figures (the short vertical line above its middle may simply be a mark to guide the engraver). At the sides, behind the outermost figures, there appear what seem to be enclosing masses of rock (more probably rock than clouds in terms of ancient conventions), probably intended as 'encircling landscape' (cf. p. 15). The two outer figures are of beardless young men who form a partly matching reciprocal pair and will be described together. They each stand with their weight on their outer leg and with their inner leg flexed, each dressed in a girdled short *chiton* with a stripe a little above its lower border and a *pilos*, either crumpled or picked out with coarse hatched shading. They have indeterminate footwear attached with thongs bound around their ankles and calves, although this lacing is indicated only on their outer legs. They both have their round shields behind them, propped against the rocks of the 'encircling landscape' which are rising almost vertically at this point; their shield devices consist of an outer circle enclosing a sort of scalloped 'sun-burst' or 'star-burst' around an inner circle (the latter detail concealed behind the left-hand figure). They also show some differences. The right-hand figure has his left arm akimbo and his right hidden. He has a swan with half-furled wings by his feet. The left-hand figure has a *chlamys* wrapped round his lowered left arm and his right arm raised in gesticulation. Next to him stands the goddess Athena (Menerva) with her right arm extended towards him and her left lowered, dressed in a *peplos* with a long, high-girdled overfold and with stripes near its edges. She has a plumed Corinthian helmet pushed up to the back of her head and wears a necklet with a round *bulla* pendant. Partly in front of her stands a nearly naked woman who looks at the youth on her right. Her garment (a *peplos*?) is completely unfastened, but gripped between her thighs, so that it still covers her right leg, and is carried around her back to where it is apparently caught under her left armpit and possibly also held loosely in her lowered left hand. She stands with her weight on her right leg and has her right arm akimbo with the hand behind her right buttock. She has laced shoes, a garland of close-set pointed leaves, bracelets and a necklet with a double-crescent centre-piece. A cord encircles her body from right shoulder to left buttock and carries three pendent cords with magical knots or amulets at their ends.

Subject. Conversation group of a type adapted by late Etruscan engravers for a variety of subjects, but with the participants here possibly identifiable by their attributes (cf. pp. 43-4 under **19**). The outer figures are probably to be recognized as the Dioskouroi on the evidence of their distinctive *piloi* and, possibly, of the 'sun-' or 'star-burst' blazons on their shields, which may symbolize the stars of the Gemini constellation (cf. R. D. DE PUMA in *LIMC* III, pp. 599 no. 10, 606). The swan also recurs between them on some Dioskouroi mirrors (*op.cit.*, pp. 600 nos. 28-30, 607, pl. 483; *CSE* France I, 2, no. 57) and is thought to be an attribute recalling the circumstances of their birth. Here, however, it is associated with only one of the pair and it is possible that the artist may be trying to distinguish between the immortal egg-born Polydeukes and his more normally born mortal brother. It is much more obscure whether this attribute is also shared by the nearly nude woman beside him, possibly marking her as his equally egg-born sister, Helen, rather than as Aphrodite, as her type might suggest; for her amulet pendants, cf. RICHTER, *MMBronzes*, no. 818. On this basis, the figures are possibly, from left to right, Castor (Castur), Athena (Menerva), Aphrodite (Turan) – or Helen (Elina) – and Polydeukes or Pollux (Pultuce).

Engraving Characteristics. The engraving is of moderate depth and firm and well-controlled. There is little internal body detail apart from the knees and ankles of the Dioskouroi and the breasts and a tiny line below the navel of Aphrodite (or Helen). The drapery folds show a degree of realism, but the hair curls display late stylizations. The eyes have an extended upper lash line with the pupil dot or the curved line indicating the iris pendent from it and with the lower lash line (and sometimes also a lower lid line) meeting it well before its outer end.

Fabric and Date. Etruscan, Spiky Garland Group, Hellenistic. Mirrors of this kind formed the later element of BEAZLEY's Class Z (*JHS* LXIX, 1949, pp. 16-17) and were more firmly linked to the Clusium Group of vases and their successors and thus to northern production centres such as Chiusi and Volterra by S. HAYNES (*MdI* VI, 1953, pp. 29-45). The Group itself was originally designated the *Kranzspiegelgruppe* by R. HERBIG (*StEtr* XXIV, 1955-6, pp. 183-205), but its English name more aptly draws attention to the *Stachelkranz* that he regarded as its distinguishing mark. For more recent views, see FISCHER-GRAF, *Vulci*, pp. 1, 112; L. BONFANTE WARREN in *GettyMusJ* VIII, 1980, pp. 147-54 and in DE GRUMMOND, *Guide*, pp. 157-60. The date suggested above is based on the present mirror's style, particularly its eye treatment, and a lowered chronology for a tomb group at Vulci (M. T. FALCONI-AMORELLI in *ArchCl* XXVIII, 1976, pp. 235-41, pls. 88-91). HERBIG, however, proposed to place the Group *c.* 100 BC and others have suggested Late Hellenistic or Roman dates (R. ENKING in *AA* 1948-9, pp. 183-237; H. SALSKOV ROBERTS in *Bronzes hellénistiques et romains* pp. 157-60; U. HÖCKMANN in *JdI* CII, 1987, pp. 247-89, especially pp. 265-76 and in *Atti 2° Congresso Internazionale Etrusco* II, Rome, 1989, pp. 713-19). Garlands of the kind decorating their borders do indeed appear in Roman art and are possibly to be equated with the Roman *corona sutilis*, but they were also known much earlier among the Greeks and Etruscans, appearing on lid sculptures of Etruscan Hellenistic cineraria (SALSKOV ROBERTS, *op.cit.*, pl. 96, figs. 5-7), while, from the 4th century BC on, the frontlets worn by Etruscan women sometimes take the form of half-garlands of this general type, as attested by surviving gold examples (e.g. HIGGINS, *Jewellery*[2] pp. 150-51, pl. 41a) and numerous representations (e.g. the New York mirror providing the title vignette of the CSE series, FISCHER-GRAF, *Vulci*, no. V51, pl. 22). Garland borders of this kind occasionally also appear on Etruscan lidded mirrors and, exceptionally, on a 4th-century BC Greek example (HAYNES, *op.cit.*, p. 42, pl. 5). The shape of these mirrors is even more characteristic of the Group, with both outer band and raised border on the *rev.* and radiating ribbing on the *obv.*; they share their *obv.* form with the main body of the Dioskouroi mirrors and thus it need not cause surprise that some typical Dioskouroi mirrors have 'spiky garland' borders (e.g. *CSE* Bologna, I no. 35).

11. Engraved mirror. Figs. 11a-d

Inv. GR.10.1972. Unprovenanced. Bought from the Cunliffe Fund and National Art-Collections Fund, 1972.

AnRepFitzwilliam, 1972, p. 8, pl. 1 and drawing on back cover; *National Art-Collections Fund Annual Report*, 1973, p. 7, pl. 12; *Country Life* CLIII, 1973, p. 759 with fig.; *The Antique Collector* XLIV, 1973, p. 107 with fig.; BROMMER, *Denkmälerlisten* III, p. 275 no. 4; FISCHER-GRAF, *Vulci*, pp. 45, 46, 70, 73-5 no. V44, 76, 83, 84, 112, 114, 115, 125, pl. 20; U. FISCHER-GRAF in *Aufnahme fremder Kultureinflüsse*, p. 176; DE GRUMMOND, *Guide*, p. 1; O. TOUCHEFEU in *LIMC* III, p. 721 no. 4, *s.v.* 'Elpenor'; D. W. J. GILL in *JHS* CX, 1990, p. 291 no. 11, pl. 6*b*.

Bronze mirror with bone handle. Original patina of mirror, to judge from surviving traces on *obv.* and edges, partly deep green malachite with small patches of light green over numerous pittings, partly bright exposed metal protected by tin oxide patina; much of heavily patinated areas of *obv.*, however, cleaned manually down to underlying cuprite layer; *rev.* exhaustively cleaned down to bare metal throughout, but has since acquired brownish natural repatination; pitted. Large centre point on *rev.* at about diaphragm of central engraved figure. On technique of mirror see Appendix. Handle lacks its end caps. Its lost internal packing replaced in modern times to provide socket for mirror tang. Surface somewhat chipped at ends and along one side; carries greenish stains from bronze.

D. 17.15 cm. H. with handle 30.9 cm, without handle 25.8 cm. H. of disc and extension 19.05 cm. H. of tang 6.75 cm. W. at bottom of extension 3.1 cm. D. of handle 3.1 cm. H. of handle 11.8 cm. Wt. with handle 606.8 gm, without handle 501.5 gm. Composition (see p. 54): copper 91.0%, tin 7.83%.

Handle. Hollow cylinder with knife-trimmed interior and lathe-turned exterior with astragal relief pattern (cf. *CSE* Denmark, 1 nos. 4, 24), whose units vary in height in somewhat random fashion, but with tallest and shortest at opposite ends. While it cannot be proven that handle belongs to mirror, its correspondence in width with bottom of extension favours its doing so.

Mirror. Circular disc with slight tendency towards ovoid at bottom where it has shallow, tapered, slightly concave-sided extension; tapered tang below for insertion in handle.

Obv. Slightly convex reflecting surface bounded by beading in relief and with edge (of projecting channelled oblique profile) ridged with half-round moulding just below and decorated in relief with tongue pattern with raised outlines, all of these motifs continuing along sides of extension. Extension has engraved decoration shown reserved against stippled background and forming complex symmetrical pattern of eight interlinked volutes sprouting six palmettes with between five and seven petals each. Lower part of engraving is of moderate depth (but less than that of *rev.*), upper part is very faint, with stippling quite obliterated; save at sides of extension, relief beading is also severely worn in ancient repolishing.

Rev. Engraved decoration of extension and border is shown reserved against stippled background. On extension and above are inverted five-petalled palmette and upright nine-petalled one in same distinctive style as *obv.* ornament, flanked by symmetrical double volutes which taper upwards as they merge into border decoration of double palmettes set back-to-back and linked by double volutes (although first palmette to left is necessarily single), border palmettes being normally nine-petalled, but eight-, ten- and eleven-petalled examples also occurring. Medallion has base-line and three-figure composition above it shows some elements of low-eye-level perspective (see below). Scene apparently set out of doors, to judge from rocky outcrop, and in front of building, marked by fluted Doric(?) column behind. The three figures are all identifiable because of a group of mirrors by a different artist, the Circe Master, employing related iconography and naming the participants (see below). To the spectator's left is Odysseus (Uthste), shown as long-haired and bearded and with his weight on his right leg. He threatens the woman in the middle with the leaf-shaped sword in his right hand which he has apparently just drawn from the round-chaped scabbard brandished on high in his left hand. He is naked apart from a generous *chlamys*, apparently folded over a third of the way from the top and fastened about his neck so that both thicknesses hang down his back, two of the corners being weighted. The woman in the middle is Circe (Cerca), shown seated with her back to him, but with her chest and head slewed round and her arms raised in alarm. Her seat is a backless chair, or *diphros*, its frame apparently through-mortised and tenoned to its turned legs and the cross-hatched thonging(?) of its top deliberately interrupted to allow a view of the upper part of its visible rear leg. She wears a thin *chiton* with buttoned sleeves revealing the outlines of her breasts and nipples and, over this, a *himation* wrapped round her hips and legs and with three weighted corners visible. Her jewellery consists of a dotted necklet and earrings apparently with large leaf-shaped plates dotted around the edges and round finials at the bottom. Her feet rest on an outcrop of rock, it being uncertain whether or not her right foot is intended as wearing a light sandal. To the spectator's right stands Elpenor (Velparun) shown as young and beardless, his right foot resting on the outcrop of rock beside Circe and his weight also supported by a vertical staff gripped with his right hand while, in his lowered left hand, he holds an unstrung bow. His only dress is a large *chlamys* worn in the same way as that of Odysseus. The garments of all three figures are shown *semé* with tiny four-dot rosettes. In addition, the *chlamydes* and *himation* have stippled borders, a notable feature in the stippling being a narrow reserved band at the bottom edge at some points where the folds overlap. Circe's *diphros* is shown in low-eye-level perspective following an inaccurate convention that seems to have evolved in Greece in the third quarter of the 5th century BC and to have persisted there into the last quarter of that century (G. M. A. RICHTER, *Perspective in Greek and Roman Art*, London, 1970, pp. 33-4, figs. 151-2; M. A. ELVIRA in *AEArq* LVI, 1983, pp. 56-7, fig. 13). In this the concealed fourth leg is ignored and the frame supporting the seat is shown as adopting a triangular instead of a four-sided shape. The same convention recurs on an Etruscan mirror in Copenhagen, of the transition from Early Classical to Classical (*CSE* Denmark 1, no. 26, dated too late). Other features of the Cambridge mirror in addition to this outmoded perspective convention evoke echoes of the 5th century BC: the head of Odysseus and the hair-style of Circe (although this is engraved with the aid of late concentric-semicircle stylizations) and, possibly, minor details, such as the treatment of her hands (cf. the Copenhagen mirror cited above).

Subject. Odysseus (Uthste) and Elpenor (Velparun) threatening Circe (Cerca) outside her palace. As mentioned, the identification is based on three Vulci mirrors, (cf. *LIMC* III, p. 721 nos. 1-3, pl. 550), showing much the same scene and naming the figures in cartouches, in New York (FISCHER-GRAF, *Vulci* no. V45), Paris (*ibid.*, no. V46) and now lost (GERHARD, *ES*, pl. 403, 2), all attributed by MANSUELLI to his Circe Master (*StEtr* XIX, 1946-7, pp. 53-4 nos. 1-3), from whose work FISCHER-GRAF has deleted a fourth mirror with a different subject in Paris (*Vulci*, no. V70), but added a mirror in Berlin (*Vulci*, no. V47). The Circe Master's versions of the scene are in a contemporary Late Classical style, omit the column of Circe's palace, have Elpenor helmeted and holding an arrow for his bow instead of a staff and introduce one of Odysseus' transformed companions in the foreground, shown as half-man, half-pig. It is to be noted that in *Odyssey* X, lines 307-47, Odysseus was alone when he confronted Circe, but that the Etruscan artists have followed a different version of the tale in which he is accompanied by Elpenor, whose main claims to fame in Homer were his youth and awkwardness and subsequent untimely death.

Engraving Characteristics. The engraved lines on the *rev.* are deep and firm on the figures and palmettes, although the latter are a little irregular; the subsidiary dotted and stippled detail is more shallowly cut. The figures are shown distinctly elongated and with mannered proportions with excessively small hands, feet and male genitals. The profile and three-quarter eyes tend to have a long, straight upper lash line (with pendent iris and pupil) extending beyond the junction with the curved lower lash line and with a subsidiary line above to mark the lid. On the male figures solid engraved lines are used for the abdominal musculature, but most other muscle detail is indicated by dotted lines, with particular attention devoted to the *serratus magnus*.

Fabric and Date. Etruscan, Early Hellenistic, possibly by an emigrant Vulcian artist. A date no earlier than this seems to be indicated by the proportions of the figures and the treatment of their eyes. FISCHER-GRAF has, however, assigned the mirror to Vulci and, on the evidence of its quality, fine hair-stranding, dotted musculature and dot-rosette garments with stippled borders, to the great age of Vulcian engraving in developed Classical times. Certainly, as the fabrics are currently identified, all these features seem to have died out at Vulci shortly afterwards, but they seem then to have spread almost immediately to various other centres and to have flourished in these on down to what would seem to be the more likely time-range of the present mirror, a few of the examples being cited in FISCHER-GRAF, *Vulci*, pp. 1-7 as 'northern' non-Vulcian products; on where they were made see also E. MANGANI in *BdA* 33-4, 1985, pp. 21-40. A possible explanation is that they were dispersed by the migration of leading Vulcian craftsmen as their own city's prosperity was increasingly restricted by the growing power of Rome, the process in that case beginning long before the Roman defeat of Vulci in 280 BC. Thus it need not cause surprise that the Cambridge mirror shows links with such 'northern' products (e.g., cf. the figure of Elpenor with that of Polydeukes, *Pultuke*, on a mirror in Florence: GERHARD, *ES*, pl. 355; DE SIMONE, *Entleh.*, I pl. 10) or that it displays a feature met in them but alien to Vulci in the stippled background to its border. The history of the border motif is equally instructive. Apparently borrowed from Late Archaic Attic vases, it was quickly adopted by early Etruscan mirrors (MAYER-PROKOP, *Griffspiegel*, nos. S4, S11; PFISTER-ROESGEN, *Spiegel*, nos. S1, S6), most notably the Vulci(?) relief mirror in the British Museum (HAYNES, *Bronzes*, no. 71), and remains attested on Vulci mirrors in the 4th century (FISCHER-GRAF, *Vulci*, no. V76 and the single-palmette version, no. V41), but then spreads to 'northern' mirrors (e.g. K. SCHUMACHER, *Grossherzogliche vereinigte Sammlungen zu Karlsruhe, Beschreibung der antiker Bronzen*, Karlsruhe,

1890, pp. 217-18 no. 233*a*, pl. 24*a*). The Cambridge mirror, however, also retains strong links with Vulci. The pose of the two outer figures recalls the simpler late Vulci mirrors, such as those originally assigned by M. DEL CHIARO to his San Francisco Group (*AJA* LIX, 1955, pp. 277-86, pls. 78-81; *Muse* XV, 1981, pp. 54-7). Further, the same original (possibly itself an earlier mirror?) that served as a model for the Cambridge *rev.* had apparently also some years earlier in Vulci inspired three mirrors by the Circe Master, although the classicizing Cambridge artist seems to have been more meticulous about rendering some of its early details.

12. Engraved mirror. Figs. 12a-d

Inv. GR.2.1976. Unprovenanced. Given by the Friends of the Fitzwilliam Museum in 1976. Formerly collections of the Vicomte de Janzé, W. H. Forman, A. H. Browne and T. Burton-Brown.

GERHARD, *ES* III, p. 8, Paralip. no. 237**, IV, pp. 49-50, pl. 317; MATTHIES, *PS* pp. 64, 90-2, 112-14, no. EIb2; L. G. ELDRIDGE in *AJA*, 1918, p. 274, fig. 12; *Christie Catalogue*, 27 April 1976, p. 39 no. 180; *AnRepFitzwilliam*, 1976 pp. 8-9, 16; D. W. J. GILL in *JHS* CX, 1990, p. 291 no. 10.

Bronze. Vestiges of original blue azurite and green malachite patina on parts of *obv.* and handle; surface cleaned manually, particularly on *rev.* of disc and extension where taken down mostly to bare metal apart from some vestiges of cuprite. Centre point on *rev.* below lip of engraved wash-basin. On technique, see Appendix.

W. 18.3 cm. H. 33.0 cm. H. of disc and extension 22.8 cm. H. of handle 10.8 cm. W. at extension tips 5.1 cm. Wt. 485.6 gm. Composition (see p. 54): copper 89.5%, tin 9.66%.

Pear-shaped disc with concave-sided extension ending in narrow points angled downwards and with handle cast in one piece with disc and ending in stylized doe's-head (or fawn's-head?) terminal.

Obv. Disc has slightly convex reflecting face bounded by relief beading, with edge, of projecting channelled oblique profile, decorated with summary tongue pattern consisting of parallel grooves rounded at top, all these motifs continuing along sides of extension. Plain edges of neck and grip of handle topped by coarser filed beading; grip decorated with cast relief of elongated stylized flower(?) ending at neck as scallop-like frill between now partly obliterated side-petals(?). Extension engraved with curly-tipped nine-petalled palmette on stalk and with side-tendrils. Engraving now consistently shallow, but this probably due to surface loss in modern cleaning, not ancient wear, and residual cuprite clogging. The crisp beading and relatively good preservation of the upper part of the palmette indicate, in fact, very limited ancient wear.

Rev. Handle has rough, concave back. Disc and extension have engraved decoration. On extension, hatched (or ribbed?) acanthus-leaf calyx sprouts trumpet-like flower from which, in turn, emerges seven-petalled palmette with five-petalled flowers in field to either side. Beginning above each of these flowers but not emerging clearly until beyond the lower part of the medallion scene there is an undulating ivy spray with berries forming a border and, between their tips near the top, a large multi-petalled flower with a small double centre. The medallion scene framed by this border has three nude figures grouped about a large pedestal wash-basin, shown in perspective apart from its plinth, which is in side elevation. A dove perches on this plinth and, to the left, a swan strains up to drink from the basin. Water pours into the basin from a lion's head spout above, probably to be understood as set in a wall behind. Beside the spout an *amphoriskos*, also shown in perspective, hangs by a cord or thong looped through its handles and apparently suspended from one side of the multi-petalled flower of the border while, to the right of the basin, there stands a tall *hydria*, also in perspective and apparently with a decorative band around its middle and a high foot. The standing figure to the spectator's left was interpreted by GERHARD and MATTHIES as a woman, but, to judge from the modelling of chest and abdomen and the shaggy hairstyle, seems more probably a male youth. He holds a strigil in his right hand and stands apparently with his weight on his concealed left leg. He wears a necklet with three pendent *bullae* and, on his left upper arm, an armlet with three similar *bullae*. His head is turned at a three-quarter angle gazing at a naked girl on the other side of the wash-basin, while his left hand is outstretched to her right arm and breast. She has her weight on her left leg and is leaning against the wash-basin, the rim of which she grips with her right hand, while her left arm is akimbo, the back of the hand resting behind her left hip. She wears a similar necklet, but with a single pointed pendant, double or triple bracelets on each wrist, multiple plain armlets on her left upper arm and a plain anklet above her right foot. An amuletic cord of the Greek *periamma* type is looped over her right shoulder and around her torso and another, over her left shoulder, meets it below her breasts. Her long hair is bound in front with a broad head-band and tied at the back of her head in a short pony-tail. Her eyes are fixed on the youth and her figure is young and slender, with small, high breasts. To the left, a further naked young woman squats on her right heel in front of the wash-basin. Her hair is wrapped in a snood or head-cloth and she wears a plain bracelet on each wrist and an earring with a pointed pendant in her ear. She is looking at her reflection in a mirror held in her left hand. Her right hand rests on her right knee holding a toilet stick (see below on function).

Subject. Bathing scene with a pair of young lovers, probably mythological characters, the 'young hero and heroine of the wash-basin'. The groups of bathers have tended to be regarded as the largest class of genre scenes of Praenestine mirrors, but this needs to be questioned since, where analogous bathers are named on Praenestine *cistae*, they are mythological figures and this may apply even more strongly to the present young lovers. The 'young hero and heroine' seem also to appear, although with quite different iconography, in other bathing scenes on Praenestine mirrors in Bologna (*CSE* Bologna I, no. 4) and possibly in Berlin (KLÜGMANN-KÖRTE, *ES*, pl. 154, the 'heroine', in that case, the kneeling figure). A possible clue to the identity of the 'hero' may be provided by an inscribed Etruscan mirror in Providence (KLÜGMANN-KÖRTE, *ES*, pl. 122; D. G. MITTEN, *Museum of Art, Rhode Island School of Design, Classical Bronzes*, Providence, 1975, pp. 132-6 no. 37), with the more formally grouped figures at the wash-basin the long-haired young Achilles (*Achle*), his mother, Thetis (*Thethis*), and the Etruscan love-goddess, *Achuvesr*. The youthful Achilles was brought up as a girl at the court of Lykomedes in Skyros, together with the latter's daughters, and the 'heroine' could be one of those daughters, Deïdameia, who fell in love with Achilles and bore his son, Neoptolemos. Or is she, perhaps less probably in view of the 'hero's' obvious youth, Briseis or Chryseis (*Creisita*), this last being named in the bathing scene that forms part of the puzzling decoration of a *cista* in the Villa Giulia (FOERST, *GPC*, no. 108) with Achilles (*Aciles*) shown nearby? Or, quite differently, is the 'hero' the young Paris (*Alixentr*), as named in the bathing scene on a *cista* in Berlin (BORDENACHE, *Ciste* I, 1, no. 9), although there, one suspects, the hero's role may be rather different (as judge in an uncanonical beauty contest) ?

Toilet sticks (σπάθαι *spathae*), as held by the squatting woman, were first identified by BEAZLEY (*EVP*, pp. 78, 111) and discussed by D. K. HILL (*Archaeology* XVIII, 1965, pp. 187-90), although the latter's interpretation that they served only as dip-sticks for perfumed oil is surely too narrow; their shapes correspond with those of contemporary Egyptian *kohl*-sticks and they were presumably, like them, also used for mixing and applying eye-paint and possibly also lip and cheek colour. For the humour of the straining swan, cf. the bathing scene on a Faliscan *stamnos* (BEAZLEY, *EVP*, pp. 81, 84, pl. 16, 2) with a heron on the basin; a duck also occurs in a Praenestine scene of this kind (FOERST, *GPC*, no. 105). Doves, as shown here, are common in such Praenestine versions, e.g. GERHARD, *ES* pl. 108; KLÜGMANN–KÖRTE, *ES* pl. 152, 2; BORDENACHE, *Ciste* I, 1, no. 24.

Engraving Characteristics. Engraved lines of *rev*. also now fairly shallow but this probably mainly due to residual patina clogging them and to heavy surface loss in modern cleaning. Lines are fluent but somewhat hasty and often made with more than one incision, with small careless overlaps. Line of left hip of youth extended too far down into area of basin and error concealed by incorporating it into squatting woman's mirror. The eyes are summarily rendered with curved upper and lower lash lines and a very short horizontal line between for the pupil. Distinctive features are the 'intermediate-profile' lines on the three-quarter head of the youth and the figures' feet and ankles.

Fabric and Date. Praenestine, Early Hellenistic. The shape is still close to that of the fine, large, Late Classical mirrors, but several features in the engraving suggest that it is somewhat later: the style and proportions of the figures, the advanced shape of the *hydria*, the late and rather debased extension florals and the way that the garland no longer forms a complete border to the scene; for the medallion's intrusion into it, as in the hanging of the *amphoriskos* from its central flower, cf. bathing-scene mirrors in Brussels (LAMBRECHTS, *Mir. Mus. Royaux*, no. 6) and Dresden (*CSE* DDR 2, no. 18). The *amphoriskos* itself is probably to be understood as of a long-current glass variety rather than the early-3rd-century silver type attested from a tomb at Bolsena (D. E. STRONG, *Greek and Roman Silver Plate*, London, 1976, p. 103, pl. 28*a*); for the way that it is hung, cf. the 3rd-century Chiusi finds (L. G. ELDRIDGE in *AJA* XXII, 1918, pp. 263-6, 272-5, fig. 7) and the fusiform unguentarium on the Barsanti *cista* in Palestrina (BORDENACHE, *Ciste* I, 1 no. 49). A mirror in Munich (GERHARD, *ES*, pl. 78) showing Leto, Artemis and Apollo and possibly still of the end of the Late Classical Period, to judge from its style, florals and garland type, may, if one can trust the reproduction, be an earlier work by the same engraver, whom we may accordingly tentatively name the Delian Triad Master. His style may have been influenced by the engraver of a more careful and possibly distinctly earlier mirror in the British Museum (inv. GR 1873.8-20.104: WALTERS, *BMBronzes*, no. 719, ADAM, *Miroirs prénestins* no. 4).

DUBITANDA IN THE FITZWILLIAM MUSEUM

The two items dealt with here were received as gifts by the Fitzwilliam Museum in 1941 and 1977, **13** originally as a forgery and **14** as of doubtful status until its decoration could be adequately revealed. Both are related to classes of objects currently regarded as spurious and they may indeed be ingenious modern forgeries, but, in view of the open verdicts returned on them by the Museum's conservators, it seems wise at this stage to leave the question of their status still somewhat open.

13. Circular relief. Figs. 13a-d

Inv. GR.1.1941. Unprovenanced. Given by R. Holland Martin, CB, in 1941.

Unpublished.

Lead. Bottom, lower right edge and small parts of relief ground elsewhere missing; some other areas crumpled or subject to bruising and abrasion, most of this damage occurring before patination. Shallow, mostly even light brown patina with darker brown patches and numerous tiny dark pittings (partly due to trapped bubbles?). Object consists of inner relief disc superimposed on plain outer ring. Inner disc cast in mould apart from background at sides beyond left and right edges of relief group which seems to have been added as thin sheet lead rather inadequately attached with small drops of tin solder. Disc then presumably retrimmed to circle and outer edge thinned by hammering. Flat outer ring cut from sheet lead with inner diameter perhaps 2 cm smaller than outer diameter of inner disc, its inner edge heavily soldered to back of disc and inevitably fusing extensively in process and securely fixing the two separately attached parts of the relief background mentioned above. No soldering to it of outer edge of inner relief disc, which instead affords a narrow slot all round between the two layers of metal penetrated by faint tool marks (as if for tucking in edges of gold leaf overlay on disc, but no surviving traces of gold).

D. 10.7 cm. D. of inner disc 9.35 cm. W. of relief group (from right elbow of Apollo to left elbow of Hermes) 8.0 cm. Wt. 105.1 gm.

Thin circular relief medallion extended in diameter by outer ring which also seems originally to have provided purchase for nails (perhaps once ten in number, with holes for six still preserved) securing it to wooden surface.

Obv. Group in low relief with all three figures identifiable both by attributes and by their names inscribed on other works showing same scene: engraved mirror formerly in Munich but lost in World War II (GERHARD, *ES*, pl. 74; I. KRAUSKOPF in *LIMC* II, p. 351 no. 119, with fig.) and a group of relief mirrors intimately related to the present example (see below). To the left sits the god Apollo (Apulu), his right hand at his side resting on the rocks on which he is seated and his left held loosely in front of his left shoulder. His garment is a *himation* wrapped around his legs with one edge draped over the rocks on which he sits (the rocks themselves not otherwise indicated save under his right hand) and also pulled up behind his back, with a corner hanging down over his left shoulder. He also wears a necklet with five pendent *bullae* and a laurel garland. In the middle is the god Zeus (Tinia) looking at Apollo and standing a little in front of the others with his weight on his right leg, grasping a round-topped sceptre in his raised right hand and loosely holding a thunderbolt in his lowered left. The thunderbolt appears to have an arrowhead at one end and a spearhead framed by a volute-and-flower motif at the other. Zeus is shown as youthful and beardless like his companions and naked apart from a necklet with pendent *bullae* and a laurel garland in his hair. Behind him stands Hermes (Turms) facing the other two with his weight on his left leg. His right hand rests on Zeus' left shoulder and his lowered left hand lightly clasps an exceptionally long *kerykeion* which rests against his left upper arm and whose top takes the form of confronted snakes. He has a winged *petasos* on his head and, to judge from the thongs indicated around his left ankle and foot, sandals on his feet. He is dressed in a variant form of the symmetrical *himation*. One end of it is draped over his right shoulder and hangs down his right side, while the rest of the garment is brought across his back and, instead of hanging over his left shoulder, passes under his left armpit and over his left upper arm and elbow to hang down his left side below his left hand.

Rev. Undecorated; almost vertical woodgrain(?) traces in patina.

Subject. Apollo (Apulu), Zeus (Tinia) and Hermes (Turms), apparently in conversation.

Style Characteristics. Cast in very shallow low relief, with a little of the detail lightly retouched after casting, a small stamp being used to add the nipples of all three figures and the navels of Apollo and Zeus, and the sandal straps on Hermes' left ankle also incised afterwards. True profile eyes, but drapery folds have rigid, flat and flaring character and formal patterning that seem classicizing with archaistic tendencies.

Fabric and Date. Probably a nineteenth-century Italian forgery, but, if so, one showing a considerable skill and prescience and whose technical characteristics probably preclude any categorical condemnation at this stage. The relatively rare Etruscan mirrors with cast reliefs on their *rev.* have often been usefully listed (e.g. DE GRUMMOND, *Guide*, p. 10 n. 9; U. FISCHER-GRAF in *AntK* XXV, 1982, p. 117 n. 3). Their technique seems to have precluded any significant ancient replication (although modern forgers have sometimes attempted to make good the omission). The four mirrors with reliefs on their backs intimately related to the Cambridge medallion have most recently been discussed by U. FISCHER-GRAF (*AntK* XXV, 1982, pp. 117-23). A silver mirror from the Karapanos collection in Athens and a bronze one in Geneva seem to carry reliefs identical with that in Cambridge (*ibid.*, pp. 120-22, pl. 22). Two further silver mirrors, the one in Florence reputedly from Bomarzo and the other in New York (*ibid.*, pp. 117-23, pls. 20-1; *LIMC* II, p. 351 no. 118), show exactly the same scene, but with the drapery hanging below Hermes' left hand slightly reduced and with the rocky outcrop on which Apollo is seated more completely added in. Although set on larger mirrors, these last two reliefs seem marginally smaller than the others which may thus have primacy in a process of mechanical copying (FISCHER-GRAF, *op.cit.*, p. 121, table 2; the comparisons in her table 3, based on photographs, are probably less reliable because of the effects of different lighting and lenses). R. NOLL accepted the Bomarzo mirror in Florence as authentic and regarded the other relief mirrors of the same type as forgeries copied from it (*ÖJh* XXVII, 1931-2, pp. 159-60 figs. 100-2; XXIX, 1935, *Beiblatt*, pp. 203-7 figs. 64-5) and this view has generally been followed since (most recently in E. PAUL, *Gefälschte Antike*, Leipzig, 1981, p. 258, pls. 212-13). But FISCHER-GRAF has argued convincingly against the authenticity of the Florence mirror also and suggested that all the relief mirrors with this scene are nineteenth-century forgeries from the same source, inspired by the engraved mirror formerly in Munich cited earlier. It is more difficult to see where the Cambridge relief fits into all this. It does not seem to represent any readily understandable technical stage in the production of the forged mirrors, but could have been made with them to strengthen their claims to antiquity by its seemingly ancient patina and possibly in response to then-current theories about a lost-lead process in ancient bronze-casting. But if, on the other hand, the Cambridge relief is to be viewed as playing any role in connection with mirrors of ancient type it would seem more likely to have been as lid decoration for a class known mainly from representations and only now beginning to be recognized – the covered mirrors in wooden boxes of Hellenistic Volterra (DE GRUMMOND, *Guide*, pp. 21-4, figs. 29-32). It thus seems desirable not yet to deny completely the possibility that it may be an ancient casting from an ancient mould, inspired by the same original as the Munich mirror-engraving but of later, possibly Hellenistic execution. In this case, a better-preserved casting from the same mould as the Cambridge relief may have served as the model for the modern relief-mirrors just discussed.

14. Engraved mirror. Figs. 14a-d

Inv. GR.3.1977. Unprovenanced. Given by Miss Gladys Davies in 1977. Formerly collection of Sir Geoffrey Taylor. Carries old painted number: P.I./59.

AnRepFitzwilliam 1977, p. 11; D. W. J. GILL in *JHS* CX, 1990, pp. 291-2 no. 12.

Bronze. Lower part of extension and handle missing; as received, part of alien ancient strigil(?) handle (now removed) had been attached instead with three modern rivets whose holes remain visible. Fine green malachite(?) patina with some blackish patches and areas of brown incrustation. As received, already partly cleaned manually, especially over extension area, but with much of engraved decoration still hidden by patina; accordingly, clearest areas (over middle part of right-hand figure) left untouched to provide control and rest of engraving cleaned out manually in Museum. Apparently not cast but cut from sheet bronze and beaten into shape with cut edge hammered up and finished by punching or crimping.

W. 12.95 cm. P.H. 13.5 cm. Wt. 61.9 gm. Composition (see p. 54): copper 85.0%, tin 9.3%, lead 4.869%.

Thin, light, pear-shaped mirror, both disc and missing handle being presumably cut in one piece from same metal sheet.

Obv. Reflecting face of fairly high convexity bounded by shallow channel and upturned cut surface decorated with crimped undulations, all these features continuing along sides of extension; edge profile approximately rounded.

Rev. Middle part engraved with a four-figure scene. To the spectator's left is a very thick-set Athena, leaning against or sitting on what appears to be not so much a seat as a couch with plain tapered legs, looped-up under-drapes, a horizontally-panelled main-frame and a curved back or primitive *fulcrum*. She is wearing a plumed helmet pushed up on to the back of her head and a *peplos* with a long girdled overfold, its open side leaving her right leg exposed, revealing that she is wearing footwear attached with thongs bound round her ankle and calf. She is apparently also wearing her *aegis* (indicated only by its scalloped edge running obliquely across her breast and right shoulder) and her left hand is raised towards her chin. Behind and to one side of Athena appears the upper part of a woman(?), otherwise concealed behind the other figures, wearing a plain necklet and with the head turned at a three-quarter angle towards the goddess. The figure may be holding a lyre whose upper part appears adjacent, unless this is carried by a strap on the back of the young man in front of her. He stands with his weight on his left leg. He is beardless and naked apart from footwear like that worn by Athena and with similar binding around the ankles and calves and, possibly, but only cursorily indicated, a meagre *chlamys* fastened at his neck and hanging down his back. His head is turned at a three-quarter angle towards the right-hand figure. This last is a beardless young man half-seated facing Athena with his ankles crossed and his left arm across his chest supporting his right elbow, his right hand held towards his chin. He is naked apart from similarly bound footwear, a *chlamys* fastened with a round brooch at his neck and hanging down over his back and right shoulder and an extremely elongated Phrygian cap shown with ridges or hatched shading. Over the surface on which he is seated is spread what may be an animal pelt with two of its legs curling up by his ankles, and he has a long-hilted sword in a scabbard propped up behind him, its strap loosened from his shoulder.

Subject. Athena (Menerva) in conversation with two young men and a woman(?). In view of their attributes, these figures could be Antiope, Amphion and Zethos. For Etruscan mirrors apparently showing these last three, see GERHARD, *ES*, pl. 219; REBUFFAT, *Miroir*, no. 45. If so, however, Antiope has been forced into the background with the introduction of Athena in the employment of a reversed version of a quite different established late Etruscan iconography. This is normally used for indeterminate conversation groups, although some of these have in the past been interpreted as a stage in the judgement of Paris. One known from a possibly inaccurate old reproduction

(D. VALERIANI, F. INGHIRAMI, G. B. VERMIGLIOLI, *Etrusco Museo Chiusino* III, Fiesole, 1832, p. 155, pl. 155; GERHARD, *ES,* pl. 258, 1) seems especially relevant in matters such as the couch behind Athena and the elongated Phrygian cap of Zethos(?); for others, see GERHARD, *ES*, pls. 193, 258,2,3, 372,1; REBUFFAT, *Miroir,* nos. 12, 34; G.A. MANSUELLI in *StEtr* XIX, 1946-7, pp. 60 (Maestro del giudizio di Paridi II, no. 8), 61 (Maestro del giudizio di Paridi IV, no. 8).

Engraving Characteristics. Moderately deeply engraved with main outlines firmly cut, if sometimes clumsily, and with some overlaps. Awkward groin lines on male figures; chest line of Amphion(?) cuts right arm of seated Zethos(?). Very summary but naturalistic handling of hair, eyes, etc., akin to late Praenestine (or modern) work with little of late Etruscan stylizations, although these are evident elsewhere, e.g. on footwear. Suggestions of perspective on couch legs, sword and scabbard.

Fabric and Date. There seem to be three main possibilities: (1) that mirror and engraving are both ancient, possibly Etruscan-influenced late Praenestine work of Hellenistic date; (2) that the mirror is ancient, but the engraving modern; (3) that both mirror and engraving are modern forgeries. The late Mr N. C. Rayner, the conservator responsible for cleaning the mirror in the Fitzwilliam Museum, was of the opinion that the patina was ancient and that the engraving was original, partly because of the way in which it underlay the patina, partly because, despite the thinness of the metal, its cutting had not distorted the *obv.* surface and was thus probably anterior to that surface's original grinding and polishing. Nevertheless, that notwithstanding and in spite of its interesting Etruscan antecedents, the engraving does not make an entirely favourable impression, in particular in the way in which it fails to fill the whole *rev.* surface of the disc. Mirrors apparently of the same shape and technique are attested in Brussels (LAMBRECHTS, *Mir. Mus. Royaux,* no. 64), and possibly in Courtrai (*CSE* Belgium 1, no. 8) and, with spurious engraving, in Bologna (*CSE* Bologna I no. 42). A thin-walled cast variety also exists, usually unengraved and showing exactly the same shape; with the published examples at the following locations it is less clear which technique is involved: Berlin (GERHARD, *ES,* pl. 22, 1, 2), Cologne (*CSE* West Germany 1, no. 24 and other examples there cited), Viterbo (EMILIOZZI, *Collezione Danielli,* no. 587) and Marzabotto (G. MUFFATTI in *StEtr* XXXVII, 1969, pp. 270-71 no. 502, pl. 57, *a*, 2, and the other examples there cited). The shape seems to have evolved from that of the small, late, 'debased' Praenestine mirrors such as **5** above. Mirrors of beaten sheet-metal, but of silver and of circular late Etruscan shape, are associated with tomb finds at Chiusi apparently of about the late 3rd and mid 2nd centuries BC: L. G. ELDRIDGE in *AJA* XXII, 1918, pp. 260-2, fig. 5; H. B. WALTERS, *Catalogue of Silver Plate, British Museum,* London, 1921, p. 7 no. 24, pl. 4. MANSUELLI (*StEtr* XV, 1941, p. 312 no. 40) regarded the Bologna example cited earlier as ancient with modern engraving and the way in which the engraving has possibly distorted the *obv.* (cf. *CSE* Bologna I, fig. 42 d) may support his view. G. SASSATELLI (*ibid.*, pp. 59-60) has suggested that all the mirrors of this shape may constitute a class of modern forgeries, in that case probably produced no later than the mid nineteenth century since the Bologna example belonged to the Palagi collection, although this seems hard to reconcile with the examples from documented finds. The fairly high lead content of the Cambridge mirror, unusual but not unparalleled in late Etruscan mirrors (see p. 53) and normal enough in other contemporary bronzework, is matched by the Brussels and Cologne examples but not that in Bologna while that in Courtrai has a more modest but still significant percentage of lead.

III THE MUSEUM OF ARCHAEOLOGY AND ANTHROPOLOGY

The archaeological collections of this Museum evolved from those assembled in the nineteenth century by the Cambridge Antiquarian Society. Portions of these were initially displayed at the Fitzwilliam Museum and in 1883 they were all made over to the University of Cambridge as the Museum of General and Local Archaeology and of Ethnology, located at first in part of the Fitzwilliam's new Museum of Classical Archaeology (see p. 49). In 1913 the collections were moved to their present premises in what is now the Faculty of Archaeology and Anthropology and renamed the University Museum of Archaeology and Ethnology, modified in the past two decades to the Museum of Archaeology and Anthropology. Problems of documentation arise with the material described here, since only three of the ten mirrors are from definitely known sources; **15** was given by Sir Hercules Read in 1922 and **16** and **17** are from the rich collection of M. F. Ransom received in various gifts and a major bequest in 1923. With these items of recorded origin, the first two digits of their inventory number represent the last two figures of the year in which they were acquired and the other digits give the object's serial number in that year; in the case of the Ransom collection, the prefix RC is added at the beginning. Undocumented items have tended to be given serial numbers with a Z-prefix. Of the two mirrors from the Ransom collection, **17** seems only later to have been identified as from this source and to have originally been listed under the same Z-number as five other examples described here (**18–22**), leading to the possible surmise that they, too, might have formed part of the Ransom collection; of these, **19** and **22** are directly linked to each other and, likewise, **20** and **21**, since they were received wrongly assembled together. Two other items sharing a different Z-number, **23** and **24**, are interrelated in having identical modern drillings to take tags. There may well have been Etruscan mirrors in the rich collection of Italian antiquities given in 1886 by Dr J. Barratt, but whether these last two examples are from his collection or that of M. F. Ransom or some quite different source it now seems impossible to tell. A further Etruscan mirror, also undocumented, from the collections of the Museum of Archaeology and Anthropology is now to be found in the Museum of Classical Archaeology (**26**; see p. 51 below). The Roman mirrors in this collection have all been described by GLENYS LLOYD-MORGAN (*ProcCambrAntSoc* LXIX, 1979, pp. 41-4).

15. Mirror. Figs. 15a-d

Inv. 22.1332. Unprovenanced. Given by Sir Hercules Read in 1922.

Unpublished.

Bronze. Green malachite patina with dark green patches and with cuprite showing through as result of limited manual cleaning. Small centre point on rev. On technique, see Appendix.

D. 15.05 cm. H. 18.05. H. of tang 3.0 cm. Wt. 237 gm. Composition (see p. 54): copper 89.5%, tin 9.34%.

Circular mirror with small tapered tang for insertion in a handle now lost.

Obv. Slightly convex reflecting face bounded by beading in relief; edge otherwise plain, of channelled vertical profile. Sharpness of beading suggests that mirror saw only limited ancient wear. Unengraved.

Rev. Back only summarily ground down in finishing and tang retains possible vestiges of rough, concave, cast surface. Unengraved.

Fabric and Date. Etruscan, Late Archaic or Early Classical. For this early shape in plain Etruscan mirrors, cf. *CSE* Bologna I, nos. 24-5, II, no. 11; for the beaded edge on such plain mirrors, cf. *CSE* Bologna I, no. 23.

16. Mirror, originally lidded (?). Figs. 16a-b

Inv. RC.23.249. From Caere; marked in ink 'Caere or Argylla, Cervetere, Italy' (the 'Pelasgian' and modern names of Caere slightly misspelt). Bequeathed by M. F. Ransom in 1923.

Unpublished.

Bronze. Lid missing and more than quarter of circumference of disc and flange corroded away including all traces of hinge. Mainly reddish brown largely cuprite surface patina with patches of green malachite and eruptions of vivid green chloride (now treated). Centre point on *rev.*

D. 11.65 cm. Depth, with flange, 1.35 cm. Wt. 203 gm. Composition (see p. 54): copper 89.5%, tin 6.41%, lead 1.35%.

Circular mirror cast as flanged disc.

Obv. Moderately convex reflecting face; lathe- or wheel-trimmed horizontal relief bands on lower part of flange, top of uppermost providing slight projection against which missing lid might have closed.

Rev. Wheel-cut relief pattern of wide and narrow concentric circles around small raised inner circle containing centre point and extending out almost to angle of flange.

Fabric and Date. Etruscan, advanced Late Classical–Hellenistic. For Etruscan mirrors of this form, see under **3** above. For layout of concentric-circle pattern of *rev.*, cf. LAMBRECHTS, *Mir. Mus. Royaux*, no. 14; *CSE* Denmark 1, no. 9.

17. Engraved mirror. Figs. 17a-d

Inv. RC.23.317 (also carries superseded earlier inv. no. Z.23214*a*). From Città della Pieve; carries sticker on *obv.* in probably nineteenth-century hand: 'Etruscan mirror found at Città della Pieve, 1856' (date either '1856' or '1556', second digit being either '8' whose upward stroke failed to register, leaving no trace also under ultraviolet light, or round-topped '5'). Bequeathed by M. F. Ransom in 1923.

Unpublished.

Bronze. Assembled from fragments, but virtually complete; heavily cracked as consequence of deep oxidation; handle bent back obliquely in antiquity (in order to prevent re-use after burial in grave?). Malachite surface patina, light green with brighter green patches, over cuprite; *rev.* of disc cleaned manually so far as practicable, but this not always helpful (in preparing fig. 17a, working photographs made before cleaning proved useful in some areas of deep cuprite accumulation). Centre point on *rev.*, by left hip of engraved figure of Paris. On technique, see Appendix.

D. 13.35 cm. H. originally 27.2 cm, now with handle bent back *c.* 26.6 cm. H. of disc and extension 16.7 cm. H. of handle 10.5 cm. W. at points of extension 3.65 cm. Wt. 285.2 gm. Composition (see p. 54): copper 86.5%, tin 12.02%.

Circular disc with concave-sided flaring extension with angular points and with handle cast in one piece with disc and decorated all round in relief. Handle ends in terminal in form of stylized head of ram with schematic horns framing ears and with slightly ribbed relief lines marking tear-ducts and eye-outlines; middle part of grip occupied by slightly ribbed torus with heavy incisions each side; similarly incised bands above and below; neck decorated on both sides with flower in relief framing bottom of extension.

Obv. Small highly convex reflecting face bounded by channel and, outside this, by radiating pattern of rounded ribs in relief extending around edge, which is of bulbous raised profile; this edge decoration also continues along sides of extension which carries schematic engraved floral(?) ornament, possibly analogous to that of **10**, but with much of it no longer traceable because of the poor condition of the surface, which also makes it impossible to assess the level of ancient wear in this area.

Rev. Concave medallion area of disc is surrounded by raised flat border zone and, raised again above this, by outer band which is prolonged along sides of extension.

Interior of extension, at intermediate raised level, engraved with stylized elongated flower (or plant?) with volutes at bottom, probably much as on **10**, but with much of it obliterated by heavy oxidation. Border zone engraved with 'contrived' garland of regularly arranged, close-packed, pointed laurel(?) leaves apparently assembled in the form of two 'sprays' bound together with a ribbon(?) at the bottom and meeting at the top (on such garlands see pp. 29-30). The medallion is engraved with a four-figure composition set in front of a temple-like building with fluted, possibly Ionic columns and with a seemingly open pediment with a central vertical support, this last and also the architrave and cornice diagonally hatched. The right-hand figure is named as *Priumne* (Priam) in the inscription above on the outer band (see below). He is apparently shown seated on a rocky outcrop; a series of round shapes, probably intended as boulders, continues on up behind him towards the roof of the building, presumably as a partial 'encircling landscape' (see p. 15). He is shown beardless, his left hand holding a staff or sceptre while his right is raised towards his mouth. He wears sandals bound with thongs about his feet and ankles, a fine helmet of Phrygian type pushed back from his face and a *himation* shown *semé* with small crosses and wrapped around his hips and legs with one corner carried up behind his back and over his left shoulder. A small area between his helmet and face and a column of the building behind is stippled to provide a contrasting background. In front of him stands a long-haired youth, named in the inscription as *Helaśntre* (presumably Alexander, i.e. Paris). He is naked apart from sandals fastened with thongs about his feet, ankles and calves and a *chlamys* looped over his left shoulder. He stands cross-legged with his weight on his right foot, steadying himself with a staff wedged under his left armpit and held with his left hand. He has his right arm bent with the hand resting behind his right buttock and his head is turned at a three-quarter angle towards Priam. Behind his right shoulder stands a woman in a long girdled garment, possibly a *peplos*, and named in the inscription as *Menerva* (Athena). She has a helmet of mixed type with a horse-hair plume pushed up on top of her head and is looking at the young man seated at the left-hand side of the composition. His name, incompletely preserved in the inscription, appears to be *Tarch....*, which has been tentatively restored below as *Tarch[unus* (Tarchon). He is half-seated on the edge of a rectangular block or chair, with his left foot raised, resting on higher ground, and his left forearm supported on his left upper leg, while his right hand is raised. He is naked apart from sandals bound with thongs about his feet and ankles, a weighted(?) *chlamys* apparently draped about his back and left shoulder and what is probably a Phrygian cap on his head. Heavy oxidation obscures some details. A seeming necklet above his right shoulder may alternatively represent the fastening of his *chlamys* or a strap for the quiver(?) and bowcase(?) behind his back.

Inscriptions. Written from right to left along raised outer band of *rev.*, beginning above head of Priam and naming persons depicted in sequence

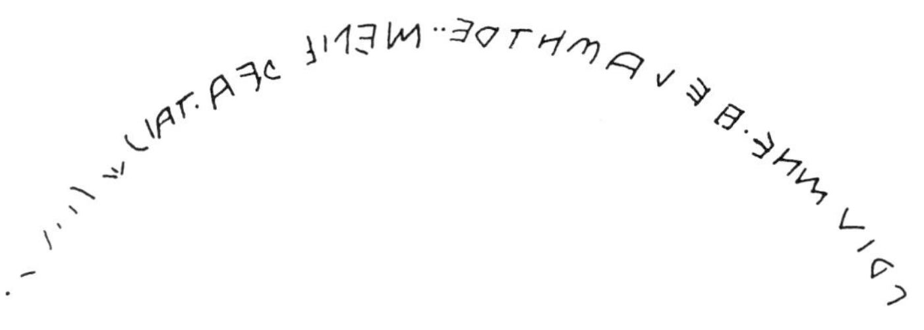

Tentatively restored:

Priumne · Helaśntre ·· Meṇẹrva · Taṛχ̣[unus] ·

Priumne is the normal Etruscan form for Priam (DE SIMONE, *Entleh.* I, pp. 102-3). *Helaśntre*, on the other hand, seems to represent a unique aspirated form of the variously spelt *Elachsantre* (Alexander, i.e. Paris) unmatched in DE SIMONE, *Entleh.* I pp. 56-7; for the omission of the χ or *c*, cf. *ibid.* no. 3, *Elsntre*. Despite some loss at its third and fourth letters, there can be little doubt about the reading, *Meṇerva* (Athena). The final name seems to have been an eight-letter word whose first two letters are certain and the next two quite likely. One might reasonably expect the name of another member of Priam's family, but there seems to be nothing in the catalogues of the sons of Priam (Apollodoros III.xii.5; Hyginus, *Fabulae* XC) that seems likely to transcribe into Etruscan to meet the requirements of the present inscription. The restoration offered here, *Tarχ[unus* (Tarchon) perfectly fits the meagre surviving indications of the remaining four letters, although this Lydian prince seems otherwise only known in Etruscan art in his post-migration phase (PFIFFIG, *Religio* pp. 37-9, 354) as the founder of Tarquinii and possibly of the whole Etruscan *dodekapolis*. The double punctuation mark between *Helaśntre* and *Menerva* may be to mark the division between the two pairs of figures. Final punctuation mark doubtful.

Subject. Conversations, probably at Troy, between Priam (*Priumne*) and Paris (*Helaśntre*) and between the goddess, Athena (*Menerva*), and, possibly, Tarchon (*Tarch[unus]*). The building in the background may have been intended to be linked with the goddess Athena standing in front of it and is possibly thus intended as her well-known temple at Troy. Since Paris is shown as a youth and Priam is not represented as aged, the scene may be set before the Trojan War in the period when Paris was first recognized by his family and brought to Troy. The identity of the left-hand figure is tantalising. If the restoration of his name suggested above is accepted then the mirror offers an interesting Hellenistic view of some sort of chronological correlation between the events just mentioned at Troy and developments in the allied territory of Lydia before the migration of the Etruscans from there to Italy under the leadership of Tyrrhenos, the father or brother of Tarchon.

Engraving Characteristics. Although now poorly preserved in some areas, the engraving appears originally to have been of moderate depth and reasonably firm and even. The drawing, though fluent, is hasty, summary and erratic. Paris's staff follows a different line above and below his left hand, the drapery folds of the lower part of Priam's *himation* acquire almost the character of a crinkly *chiton* and the architrave of the building behind follows an undulating line, with the columns set at different angles. The profile and three-quarter eyes simply have curved upper and lower lash lines meeting at one corner and with the pupil, where still detectable, a tiny dot just above the lower lash line.

Fabric and Date. Etruscan, Spiky Garland Group, Hellenistic. The Spiky Garland Group has already been discussed under **10** (see pp. 30-1). In view of its simpler garland lacking *bakchos*-rings and its more debased style, this may be a later example than **10**.

18. Engraved mirror. Figs. 18a-d

Inv. Z.23214*b* (also wrongly renumbered Z.39534). Unprovenanced. Source unrecorded (possibly Ransom collection; see p. 39).

Unpublished.

Bronze. Malachite patina, light green with deeper green patches, over cuprite and brownish black cupric oxide(?), but with considerable areas of *obv.* still showing as bright metal protected by tin oxide patina; heavy corrosion only near rim on *rev.*; on *rev.* much of disc and extension cleaned to bare metal or cuprite layer. Small centre point preserved on *rev.*, under navel of middle engraved figure. On technique, see Appendix.

D. 14.6 cm. H. 30.5 cm. H. of disc and extension 17.8 cm. H. of handle 12.9 cm. W. at points 3.0 cm. Wt. 269.5 gm. Composition (see p. 54): copper 86.5%, tin 10.59%.

Circular disc with concave-sided extension flaring out at points and with handle cast in one piece with disc and ending in terminal in form of slender stylized head of doe (or fawn?) with back-swept ears, incised folds of skin across back of neck and incised star (or tuft) on bridge of nose; light, slender grip of handle decorated all round with low knobs, as if to suggest trunk of palm or other tree with vestiges of lopped-off branches (or roughly shaped club).

Obv. Disc has slightly convex reflecting face bounded by shallow channel and vestiges of relief beading, both these last also extending along sides of extension. Prolonged ancient polishing seems to have worn away most of the beading. Plain edge of slightly offset curved profile. No traces of engraved decoration on extension.

Rev. Medallion area surrounded by raised flat border zone and surface of extension set slightly higher again. Extension engraved with elongated flower(?) with ribbed petals and with forked stigma(?) of pistil protruding at top. Border zone engraved with two rather stylized sprays, presumably of laurel, forming garland, their stems meeting exactly at top and only slightly separated at bottom, but with no binding indicated. Medallion engraved with group of five figures, with feet of three foremost extending into border zone and slightly interrupting garland. These figures are apparently standing in front of a building, part of whose cornice appears behind their heads, decorated with two zones of triangles or zigzags and an upper one of tongue pattern, separated by plain bands. The composition is conceived symmetrically, with the outermost figures almost mirror images of each other. They are of nearly naked women with their weight on their inner legs gazing intently at the central figure of the composition. Their outer arms hang by their sides while their inner hands are raised towards their chins in gesticulation. These raised hands may also be intended as loosely holding the edge of their garments, which are also gripped between their thighs and carried up behind their backs, giving a veil-like effect. This last feature is rather more adequately indicated behind the left-hand woman. Too much space has been left behind her counterpart to the right and the artist has employed a wavy line as filling ornament to conceal this. Both women wear shoes, although these are only adequately indicated on their outer feet. Their other details vary more. The left-hand woman wears a tall frontlet of Etruscan type decorated with three discs and has a plain necklet and a plain bracelet on her left wrist. The headdress of the right-hand woman is less certain; it may be a rather flattened Phrygian cap. She wears a plain necklet and a torque (unless the appearance of the latter is produced by overlapping graduated discs hanging from the necklet). Beyond the left woman appear the head and shoulder of a further figure facing left. This is probably also a woman, as she appears to be wearing a similar frontlet in her hair. In the centre of the scene is a beardless young man standing with his weight on his right leg, his body turned at a three-quarter angle towards the right-hand woman while his head is turned back to look at the one to the left. His right arm is akimbo, with the hand resting a little below his hip, while his left hand is raised. He is naked save for a *chlamys* looped over his left shoulder and hanging down his back, a bracelet on his left wrist and a Phrygian cap on his head, its leather indicated with curved pock-marks and its flaps trailing behind. Behind, between him and the right-hand woman, is the figure of a further woman dressed in a long garment, possibly a *peplos*, with her head turned at a three-quarter angle towards him. Her head-dress is somewhat confused with the tongue-pattern of the cornice, but it could be a helmet pushed to the back of her head, the only elements visible at this angle being in that case the two raised cheek-pieces.

Subject. The scene broadly corresponds with the judgement of Paris, but introduces, as a fifth figure, not Hermes but a fourth woman. In other types of scenes showing Paris (e.g. GERHARD, *ES*, pls. 189, 191-3, 372,1), GERHARD has suggested that one of the women may be Paris's mistress, the nymph Oinone, apparently the Etruscan Enuna, and she might be the fourth woman here, since here, as in **4**, the contest takes place in an architectural setting, possibly that of Paris' own home. This fourth woman is absent from other versions of the scene by the same artist and her presence here may be partly due to considerations of symmetry. On this basis the figures may possibly be identified as, from left to right, Aphrodite (Turan), Oinone? (Enuna), Paris (Elachsantre), Athena (Menerva) and Hera (Uni).

Engraving Characteristics. The engraving is now of barely moderate depth, but the outlines are firm and well-controlled, with only rare overlaps, as at Aphrodite's right hand. The laying out of the design is, however, more seriously at fault at some points, e.g. in the way that the upper and lower parts of the inner legs of the outermost goddesses fail to follow through. The profile eyes have a short lower lash line meeting a lengthened upper lash that runs closely parallel with the brow and that has a large, more or less semicircular, pendent iris. The three-quarter eyes are simply outlined, but to better effect.

The artist's profile heads are quite distinctive. The hands tend to conceal either the thumb or the little finger and Paris' right foot has only four toes. Whereas the musculature of the goddesses is indicated a little crudely with solid lines, fine body hatching is used for that of Paris, the distinction recalling a considerably earlier stage in the evolution of Greek painting when light and shadow were already rendered on male flesh but its female equivalent was still shown in flat white with linear detail (A. RUMPF in *JdI* XLIX, 1934, pp. 6-23; A. RUMPF *Malerei und Zeichnung, Handbuch der Archaologie* IV, 1, Munich, 1953, pp. 120-51).

Fabric and Date. Etruscan, Hellenistic. Two somewhat similar mirrors showing analogous scenes (at least one of them apparently by the same hand) occurred in the Calini (Calisna) Sepus tomb near Monteriggioni (R. BIANCHI BANDINELLI in *StEtr* II, 1928, pp. 161-2 nos. 173-4, pl. 35), with finds from the 3rd century on. The engraving is the work of a humble but lively and highly individual artist. So as not to add further to MANSUELLI's numerous Judgement of Paris Masters (*StEtr* XIX, 1946-7, pp. 59-62), it is proposed to name him the Pondering Paris Master. In addition to the Cambridge mirror, the following seem to be by him: (1) GERHARD, *ES*, pl. 262, 2 (interior engraving of lidded mirror from Vulci, *ibid*., pl. 243, 1); (2) GERHARD, *ES*, pl. 185 (on similar mirror?); (3) BIANCHI BANDINELLI, *loc.cit*., no. 173, pl. 35 (Terrosi collection, from Tomb of Calini Sepus). Various other mirrors seem near him, if not actually by him, e.g.: (1) KLÜGMANN–KÖRTE, *ES*, pl. 102, 1; (2) GERHARD, *ES*, pl. 274, 1 (in Vienna); (3) LAMBRECHTS, *Mir. Mus. Royaux*, no. 22; (4) EMILIOZZI, *Collezione Danielli*, pp. 55-6, pl. 32 (Mursana Tomb XX, no. 1); (5) BIANCHI BANDINELLI *loc.cit*., no. 174; *CSE* DDR 1, no. 44 (Berlin inv. M.I. 10348; from Calini Sepus Tomb); (6) *CSE* DDR 1, no. 47 (Berlin inv. 33091); (7) *CSE* W. Germany, 2 no. 19 (Hamburg, private collection); (8) *CSE* France I, 1, no. 23 (Louvre inv. 1762). These shade off into a much broader group of late mirror engravings. On their late date, cf. U. HÖCKMANN in *JdI* CII, 1987, pp. 247-89, especially pp. 256-9 and figs. 1-2. For the earlier kind of work from which they probably drew their inspiration, cf. G. QUATTROCCHI, *Il Museo Archeologico Prenestino*, Rome, 1956, p. 39 no. 95, fig. 41.

19. Engraved mirror. Figs. 19a-d

Inv. Z.23214*c*. Unprovenanced. Source unrecorded (possibly Ransom collection; see p. 39). Acquired wrongly assembled with **22**.

Unpublished.

Bronze. Lower part of extension and handle missing. Surviving patina (mainly on *obv*.) clearly even originally quite thin and largely brownish with patches of green malachite and cuprite and also of bright metal protected by tin oxide patina; *rev*. heavily cleaned manually and shows some pitting. Faint patina-filled centre point on *rev*. below right hip of engraved nude young man. Rev. extension filed down in modern times to allow soldering on of **22** (now removed). Disc cracked at side. On technique, see Appendix.

D. 12.2 cm. P.H. 13.15 cm. Wt. 113.4 gm. Composition (see p. 54): copper 85.0%, tin 13.51%.

Circular disc and, originally, concave-sided extension and with missing handle probably cast in one piece with disc.

Obv. Small, highly convex reflecting face bounded by relief beading and half-round moulding flanked by slight channels and radiating pattern of ribs in relief which extend around edge, all these motifs except beading continuing along sides of extension. Edge of bulbous raised profile. What survives of extension engraved with apex of highly stylized floral(?) motif, its tip flanked by dots. Both engraving and relief beading show little ancient wear.

Rev. Concave medallion area surrounded by raised and angled outer band which continues along sides of extension (partly now obliterated by modern filing). Interior of extension set at intermediate level and showing no traces of engraved decoration. Medallion engraved with four figures standing on short base-line which is cut by one of their feet. Three almost horizontal lines, possibly suggesting the cornice of a small building behind, appear between the heads of the two right-hand figures. The outer figures form a matching symmetrical pair, probably of beardless young men, each leaning against an altar shown in rudimentary perspective and with volute-shaped supports under its spreading top. They each stand with their weight on their outer leg and with their outer arm akimbo and resting against the altar behind and they are each dressed in a short sleeveless *chiton* with a high-girdled overfold, sandals bound with thonging up to mid-calf and a Phrygian cap with a band tied round the brow holding the side-flaps tucked up. The lower part of each *chiton*, also shown in perspective, has a broad border. The left-hand figure also has his inner arm resting on the altar and perhaps also wrapped in a *chlamys* (unless overlaid by the three continuing horizontal lines of the altar); he is in conversation with a woman holding out her right arm and wearing a long garment, possibly a *peplos* with a high-girdled overfold and possibly with a similar border, as well as a rather taller but similarly bound Phrygian cap. Beside the right-hand figure is a curly-haired young man, naked apart from sandals bound with thongs to mid-calf and standing with his right arm akimbo and with his body turned at a three-quarter angle towards the middle of the composition, but with his head turned back, also at a three-quarter angle, to look at this companion, whose body obscures his left arm.

Subject. Group of three young gods or heroes and a woman or goddess, the matching outer figures with their matching altars being possibly intended as the Dioskouroi. Modern convention tends to regard the figures in uninscribed conversation groups of this class as unidentifiable because of the widely conflicting evidence of the examples where seemingly identical figures are variously named in accompanying inscriptions. The cast shape of the mirrors of the class, however, is almost identical to that of the Dioskouroi mirrors (see below) normally showing the two Dioskouroi in much the same pose and dress as the present two outer figures and it thus remains difficult, in the absence of inscriptions to the contrary, to dissociate them completely from the Dioskouroi. For a similar but finer conversation group with the participants more adequately identified by attributes, see **10** above.

Engraving Characteristics. The engraving is for the most part moderately deep, firm and bold, but there is some variation and some features are quite lightly incised. The lines themselves are frequently broken and inaccurately rejoined and, in many minor instances, are misplaced or extended too far. Harsh, largely straight lines are used for muscle detail. The eyes tend still to have the upper lash line overlap the lower, but the profile heads have virtually degenerated to the stage of having full-face leaf-shaped eyes.

Fabric and Date. Etruscan, Hellenistic. For the way in which the shape of the mirrors of this class and, in particular, their distinctive *obv.* link them closely with the Dioskouroi mirrors and, indirectly, with the Spiky Garland Group cf. pp. 30-1; for the Dioskouroi mirrors, REBUFFAT, *Miroir*, pp. 483-4, 577-92; *CSE* Bologna I, pp. 52-3; R. D. DE PUMA in *LIMC* III, pp. 597-609, especially nos. 1-7, 43-6 (for type of present mirror, cf. *ibid.*, nos. 55-6); for their late dating, U. HÖCKMANN in *JdI* CII, 1987, pp. 247-89, especially pp. 259-60. The iconography of the engraved scene recurs frequently, e.g. GERHARD, *ES* pls. 259,3, 263,6, 266,1-6, 267,1,4; *CSE* France 1, I, no. 34; 1, II, no. 47 (both by the same hand as the Cambridge mirror); and other similar examples (sometimes with a nude female figure instead of the nude male) occur on other related classes of mirrors with which at least part of the present series must have overlapped in time, e.g. *'raised-border' mirrors with laurel garland*: RICHTER, *MMBronzes*, no. 820; REBUFFAT, *Miroir*, no. 70; *CSE* France I, 1, no. 32; *'raised-border' mirrors with cable pattern*: REBUFFAT, *Miroir*, no. 23; E. REEDER WILLIAMS, *The Archaeological Collection of the Johns Hopkins University*, Baltimore and London, 1984, no. 52 (especially close); *CSE* USA 1, no. 33 (by same hand); British Museum, GR 1814.7-4.925; *CSE* W. Germany 2, nos. 12, 32 (both near); *CSE* France I, 2, no. 43 (very near); *Spiky-Garland-Group mirrors*: *CSE* USA 1, no. 24; REBUFFAT, *Miroir*, nos. 36, 69 (with shields instead of altars); RICHTER, *MMBronzes*, no. 819 (with nude female). Also close to the Cambridge scene is the Bastis collection mirror (*Antiquities from the Collection of Christos G. Bastis*, New York, 1987, no. 128). The modest artist of the Cambridge mirror may be named the Skew-Neck Master. The works near him shade off into a much wider class of late engraving.

20. Engraved mirror. Figs. 20a-d

Inv. Z.23214*d*. Unprovenanced. Source unrecorded (possibly Ransom collection; see p. 39); there are faint, indecipherable vestiges of modern ink inscriptions on two parts of *rev*. Acquired wrongly assembled with **21**.

Unpublished.

Bronze. Handle and most of extension missing. Much of surface on both sides preserved as bright metal by tin oxide patina; elsewhere patina largely dark green malachite, but also with patches of azurite and of exposed cuprite. Centre point preserved on *rev.*, by left groin of engraved Lasa. Modern blackening of surface near extension to mask transition when **21** (now removed) was soldered on. On technique, see Appendix.

D. 12.3 cm. P.H. 12.65 cm. Wt. 104.6 gm. Composition (see p. 54): copper 83.0%, tin 12.65%.

Small circular disc, its missing extension probably originally concave-sided and its missing handle probably cast in one piece with disc.

Obv. Highly convex reflecting surface bounded by channel and by outer ridge carrying relief beading (latter now rendered somewhat indistinct by ancient wear). Modern filing before soldering on of **21** has obliterated any vestiges there may have been of engraved decoration on surviving upper part of extension. Plain edge of offset oblique profile.

Rev. What little survives of upper part of extension has surface set slightly higher than concave back of disc and is engraved with upper part of inverted triangle (to judge from Berlin examples cited below, when complete probably X-shaped motif closed by horizontal lines above and below); from this there emerges what may be intended as a schematic hemispherical flower engraved immediately above on the lower part of the medallion with a winged naked woman apparently shown standing on top of it. She has her weight on her right leg, her right hand on her hip and her left arm hanging at her side. Her wings are half-spread and she wears shoes, a plain necklet and a Phrygian cap.

Subject. Lasa, probably standing on a flower. For mirror medallions with Lasa figures in different styles standing on the flowers decorating the extensions below, cf. GERHARD, *ES*, pls. 34,1,4, 35,1; LAMBRECHTS, *Mir. Mus. Royaux*, no. 53; *CSE* Denmark 1, no. 7; *CSE* Bologna I, no. 33.

Engraving Characteristics. The engraving is mostly of moderate depth, but there are considerable variations in this, with some lines quite lightly incised. The outlines are tolerably controlled but the drawing is coarse, the left arm and hand being particularly gross and the wings sprouting much too low down. The eye in the Lasa's profile head has virtually reverted to a leaf-shaped full-face variety.

Fabric and Date. Etruscan, Hellenistic. On such Lasa mirrors, see pp. 28-9. The unaspiring artist may be named the Empty-handed Lasa Master, his work being also attested from the Calini Sepus Tomb near Monteriggioni (cf. p. 43). Also by him are: (1) Berlin, Staatliche Museen inv. M.I. 10343. Lasa. BIANCHI BANDINELLI in *StEtr* II, 1928, p. 162 no. 180; *CSE* DDR 1, no. 39. From Calini (Calisna) Sepus Tomb. (2) Berlin, Staatliche Museen inv M.I. 10344. Winged youth. BIANCHI BANDINELLI *op.cit.*, p. 163 no. 184(?); *CSE* DDR 1, no. 40. Provenance as last. (3) Brunswick, Herzog-Anton-Ulwich-Museum inv. AB 106. Lasa. *CSE* W. Germany 2, no. 1. Many other mirrors are near him: (1-2) Winged youths. GERHARD, *ES*, pl. 31,4,5; further similar examples cited GERHARD, *ES* III, p. 244. (3) Lasa. GERHARD, *ES*, pl. 32,6. (4) Winged youth. REBUFFAT, *Miroir*, no. 58. (5-6) Lasae. LAMBRECHTS, *Mir. Mus. Royaux*, nos. 50, 51. (7) Winged youth. *CSE* USA 1, no. 10.

21. Handle from mirror. Figs. 21a-e

Inv. Z.23214*e*. Unprovenanced. Source unrecorded (possibly Ransom collection; see p. 39). Acquired wrongly assembled with **20**.

Unpublished.

Bronze. Disc missing. Lustrous black sulphide(?) patina with patches of green malachite. Upper part of extension cut and filed down to fit to **20** (now removed) and parts of surface in vicinity cleaned to bright metal to mask transition. On technique, see Appendix.

P.H. 15.15 cm. W. at points 2.9 cm. Wt. 68.4 gm. Composition (see p. 54): copper 86.0%, tin 12.73%.

Handle (originally cast in one piece with its missing disc) ending in terminal in form of head of doe (or fawn?) with incisions across bridge of nose and, as if representing wrinkles, across back of neck and with punched dots around edges of laid-back ears and, on *obv.*, separating *protome* from grip. Grip of rectangular section with cast(?) longitudinal channel along middle of each of four sides and with cold-tooled(?) oblique incisions, particularly near edges and with similar boldly incised but summary floral(?) motif on both *obv.* and *rev.* at neck. Concave-sided extension.

Obv. Unengraved extension bounded at sides by channel and relief beading and with plain vertical edge.

Rev. Plain, flat underside to terminal. Extension engraved with lower part of stylized elongated flower(?) with some oblique veining and summary volutes(?) at bottom.

Fabric and Date. Etruscan, Hellenistic. The form of the handle and extension and the engraved floral decoration of the latter all suggest that this is from an Etruscan 'raised-border' mirror of Hellenistic date; especially close seems *CSE* Denmark 1, no. 1; cf. also LAMBRECHTS, *Mir. Mus. Royaux*, nos. 3, 39; *CSE* Netherlands, nos. 4, 9, 24.

22. Handle from mirror. Figs. 22a-e

Inv. Z.23214*f*. Unprovenanced. Source unrecorded (possibly Ransom collection; see p. 39). Acquired wrongly assembled with **19**.

Unpublished.

Bronze. Disc missing. Heavy dark green malachite patina with patches of exposed cuprite. Rev. filed down at top for soldering to **19** (now removed). On technique, see Appendix.

P.H. 10.7 cm. W. at points 2.65 cm. Wt. 29.9 gm. Not available for bronze analysis.

Handle (originally cast in one piece with its missing disc) ending in terminal in form of stylized animal head, possibly of ram with horns encircling its large ears; also preserves part of flaring, concave-sided extension.

Obv. Middle part of grip of handle occupied by narrow torus flanked by wider tori with oblique incisions marking off outer triangular side-panels. Upper part of grip takes form of elaborate flower in cast relief, its long crinkly petals framing elongated neck of handle. Sides of extension marked by relief beading and with plain vertical edges. Extension engraved with stylized flower(?) with ribbed centre-petal(?) and outer petals(?) in profile, both accompanied by rows of dots; above 'outer petals' deeper incisions continue along sides of extension just inside beading.

Rev. Rough concave back to handle, probably slightly modified by hammering. No traces of engraving on extension.

Fabric and Date. Etruscan, Hellenistic. In view of its shape, small size and the engraved motif (cf. **9**) and beading of the extension, this should be the handle from a developed Lasa mirror; cf. LAMBRECHTS, *Mir. Mus. Royaux*, nos. 48-50; *CSE* Denmark 1, no. 19; *CSE* Bologna I, nos. 32, 33; *CSE* Netherlands, no. 19. Although this seems nearly certain, a small element of doubt must remain because of the small number of aberrant and even hybrid forms occurring in the large classes of late Etruscan mirrors such as the developed Lasa and Dioskouroi mirrors. On the Lasa mirrors, see pp. 28-9.

23. Mirror. Figs. 23 a-c

Inv. Z.23215*a*. Unprovenanced. Source unrecorded, but seemingly from same collection as **24** (see p. 39).

Unpublished.

Bronze. Smooth, even, pale green malachite(?) patina with isolated patches of bright green and of exposed cuprite; on *rev.* patina faintly preserves outline of larger curved object once in contact with mirror in soil. End of tang apparently broken away where weakened by modern drill hole and second hole of identical diameter drilled higher up to take modern copper wire presumably once attaching tag (latter now lost); cf. **24**. On technique, see Appendix.

D. 14.9 cm. P.H. 18.4 cm. H. of disc and extension 15.95 cm. W. at points 1.75 cm. Wt. 508.8 gm. Composition (see p. 54): copper 87.0%, tin 10.31%.

Circular mirror, its outline curving out below to form a small extension and with part of a tapered tang below this for securing it into a now lost handle.

Obv. Reflecting face of only negligible convexity bounded by relief beading showing heavy ancient wear and continuing half-way along sides of extension. Edge of offset oblique profile decorated with tongue pattern with relief outlines set below beading just described. Extension and lower part of disc engraved with antithetical pair of double volutes supporting seven-petalled palmette, with some of lines awkwardly broken and showing preliminary marking for longer lower petals on palmette. Lower part of engraving fairly deeply cut but upper part worn faint.

Rev. Unengraved.

Fabric and Date. Etruscan, Early Classical or the start of Classical. Cf. *CSE* Bologna I, no. 26, which, however, shows more advanced palmette decoration.

24. Engraved mirror. Figs. 24a-d

Inv. Z.23215*b*. Unprovenanced. Source unrecorded, but seemingly from same collection as **23** (see p. 39).

Unpublished.

Bronze. Handle and lower part of extension missing; disc cracked at edge. Light green malachite patina with light blue azurite patches and with areas of overlying blackish and brownish accretion on *obv.*, heavy corrosion at edge and deep pittings on both *obv.* and *rev.*; limited manual cleaning on *rev.* and lower part of *obv.*, at some points penetrating to underlying cuprite and bare metal. Faint vestigial centre point on *rev.*, on *himation*-wrapped abdomen of engraved figure with spear. Traces of solder on break at extension testify to attempted modern repair before handle lost; surviving part of extension subsequently drilled with modern hole of same diameter as those on **23**. On technique, see Appendix.

D. 13.0 cm. P.H. 14.0 cm. Wt. 166.0 gm. Composition (see p. 54): copper 87.5%, tin 10.55%.

Circular mirror with concave-sided extension and with missing handle probably cast in one piece with disc.

Obv. Moderately convex reflecting face bounded by channel and plain low ridge, both of which continue along sides of extension. Plain edge of offset oblique profile. What survives of extension apparently unengraved.

Rev. Medallion area surrounded by raised flat border zone, itself surrounded by what virtually amounts to angled outer band, this last also continuing along sides of extension. Interior of extension on same plane as border zone and engraved with upper part of daffodil-like flower with open lower petals and central trumpet-like member with curled-over lip, internal lines marking both petals and trumpet as veined or crinkled. Border zone engraved with somewhat schematic garland, apparently of laurel, the stems of its two sprays being drawn as one continuous circle. Binding of garland possibly rudimentarily indicated at bottom by device resembling three-petalled palmette. Medallion engraved with four figures, some of their feet extending slightly into border zone. To the spectator's left stands a partly naked woman, her legs together with her tumbled unfastened garment (apparently a *peplos*) gripped between them. Her left hand is also pulling one upper corner of it up behind her left shoulder while the other upper corner seems to be held in her lowered right hand. Part of it seems to be shown as caught by the breeze and fluttering behind her back. She apparently wears a plain necklet, pairs of bracelets on each wrist and boots bound with thongs about the ankles. Three beardless young men are facing her, the first of them standing frontally with his head turned towards her and his weight on his right leg. His raised right hand is apparently to be understood as gripping his vertical spear just below its head (see below), and on his left arm are poorly drawn muscles, like those on the upper left arm of the figure far right. He is dressed in a *himation* which is wrapped around his hips and legs, with a substantial overfold hanging down in front, but, behind, its surplus material is used to cover his back and is pinned across his upper chest below his throat and it also hangs down over his left upper arm. He wears shoes bound with thongs around the instep. Of the second young man little is shown save the head and possibly folds of the garment around his legs (but these folds could, less probably, belong to the drapery of the figure on the right). His hair is bound with a head-band and his neck angle shows that he is in a half-crouched posture like the figure to the right. This right-hand young man also has a *himation* wrapped around his hips and legs, and perhaps a bowcase carried on a strap over his left shoulder, but the detail is not completely clear. Both his legs are slightly bent and he is hunched forward gripping a staff or weapon in his left hand and fingering its tip with his right. He has similar shoes on his feet, bound with thongs around the instep.

Subject. Apparently a mythological scene not able to be matched among published Etruscan mirrors or in other Etruscan art. On the other hand, within its very considerable stylistic limitations, it accords to some extent with *Iliad* VI, lines 312-69, when Hector returns from the hard-fought battle to Troy, enters Paris' house still carrying his long spear and, finding him idly attending to his armour and bow (perhaps the figure on the right), rouses him to join the fight and when Helen seeks to placate Hector by taking upon herself the blame for the war. On this basis, the figures may possibly be identified as, perhaps, from left to right, Helen (Elina, variously spelt in Etruscan), Hector (Ectur), Paris' groom or squire, and Paris himself (Elachsantre, also variously spelt), but no certainty is possible.

Engraving Characteristics. The engraving is of moderate depth and the outlines are normally well controlled and firm, with awkward overlaps only rarely occurring, notably on the bowcase(?) of Paris(?) and the right heel of Hector(?). There do, however, seem to have been significant changes made in the course of engraving. Thus the left leg of Helen(?) seems originally to have been intended as bent back at the knee to compensate for the obliquity of her right leg and her shoe-clad left foot was at first engraved at the edge of the medallion behind her right foot and remains there imperfectly disguised as part of the drapery. Also Hector(?) seems originally to have been engraved without his spear and with his right hand simply raised in gesticulation, index finger extended. The spear was added subsequently but, by then, in order to be visible, its shaft had to be edged beyond the left hand of Helen(?) and the drapery it holds, thus according ill with the position of Hector's(?) own hand. The profile eyes show a shorter lower lash line meeting a longer upper lash line which extends well beyond the junction and has a semicircular iris pendant from it. The hair locks are marked by careful stranding, although heavy corrosion has destroyed most of this detail on the head of Helen(?). The figures are ill-proportioned with over-long arms. The treatment of the drapery folds varies between a certain realism on the *himation* of Paris(?) and mannered stylization on the *peplos* of Helen(?).

Fabric and Date. Etruscan (north Etruscan?), Hellenistic. Although the form of the mirror and the advanced engraving conventions leave little doubt as to the date and much in the decoration seems northern, the proportions and style remain somewhat idiosyncratic.

IV THE MUSEUM OF CLASSICAL ARCHAEOLOGY

The Museum of Classical Archaeology was originally created in 1883 as a special department of the Fitzwilliam Museum to house its large collection of plaster casts of Greek and Roman sculpture, provide a library and other facilities for the teaching of classical art and archaeology and accommodate the collections of the Cambridge Antiquarian Society, which passed under University control at this date and remained there until 1913 (see p. 39). It achieved autonomy from the Fitzwilliam Museum at the end of 1911 and in 1982 it moved to new premises which it shares jointly with the University Faculty of Classics. Its collection of antiquities has been formed as an aid to teaching, consisting partly of material transferred from the Fitzwilliam Museum and from the Museum of Archaeology and Anthropology and partly of various private gifts. The objects have been given numbers with a CAM-prefix, but these serials, largely assigned retrospectively, bear limited relation to the sequence of acquisition. The only Etruscan mirror of recorded source was transferred from the Museum of Archaeology and Anthropology at an unspecified date, but seemingly in the period when it was known as the Museum of Archaeology and Ethnology (**26** below).

25. Engraved mirror. Figs. 25a-d

Inv. CAM.343. Unprovenanced. Source and acquisition date unrecorded.

Unpublished.

Bronze. Two areas of disc missing; broken-away section of rim adjoining one of these re-attached. Disc buckled and still distorted despite modern straightening. Considerably pitted. Harshly cleaned throughout to bare metal, but showing slight subsequent natural brownish repatination despite modern seal; on *rev.* cleaning has removed large areas of engraved surface, which seems to have separated into layers under corrosion; X-rays, kindly prepared by Dr M. Frost at the Welding Institute at Abington Hall near Cambridge, have revealed residual traces of some of this lost engraving (unfortunately too faint for plates to be usefully reproduced here); only such lines as are quite certain have been added on fig. 25a. On technique, see Appendix.

D. 15.9 cm. H. 22.0 cm. H. of disc and extension 17.8 cm. W. at points 2.65 cm. H. of tang 4.2 cm. Wt. 171 gm. Composition (see p. 54): copper 91.5%, tin 6.44%.

Circular mirror, its outline curving out at bottom into slightly concave-sided tapering extension with rectangular tang below to secure it in its missing handle.

Obv. Reflecting surface probably originally only slightly convex (curve now distorted by buckling and straightening); bounded by beading in relief which also continues most of way along sides of extension. Plain edge of slightly offset oblique profile. Extension and lower part of disc engraved with antithetical pair of double volutes enclosing asymmetrical four-petalled palmette and, above, supporting nine-petalled palmette flanked by two pairs of horn-like tendrils. Engraving of upper part of this decoration only slightly fainter than lower part.

Rev. Extension apparently engraved with double-volute and palmette motif analogous to that on *obv.* but probably without horn-like side-tendrils, but only parts of this are now preserved. By each side of its upper palmette there emerges a thick but tapering stem of ivy with alternating leaves and clusters of three berries (at least three of clusters with dots at centre of berries), each encircling half the disc to form the border garland. Within this the medallion is engraved with two figures in combat, both incompletely preserved. The figure on the left has his feet firmly planted on the rocky ground, his oblique right leg taut and his left flexed, and his upper body bent forward with a knotty club held horizontally in his lowered right hand. He is apparently naked apart from a lionskin of which the fore-paws are knotted on his chest and the hind leg part emerges from behind him to be gripped in his left hand. The figure on the right is apparently on the point of fleeing, the weight momentarily on the advanced left leg, and the right hand raised while on the left arm is a round shield. The garment worn is a short *chiton* with a band at its lower border, buttoned above along the arms to form sleeves. Nothing survives of the head save the hair at the nape of the neck and, possibly, the bottom edge of the neck-guard of a helmet. A small double-axe with the shaft broken has been discarded below to the right.

Subject. Herakles (Hercle) about to slay the Amazon queen, Hippolyta (Hephleta or Heplenta). For the subject, see G.A. MANSUELLI in *StEtr* XX, 1948-9, p. 76; G. CAMPOREALE in *StEtr* XXVII, 1959, p. 122; BROMMER, *Denkmälerlisten* I, p. 21. Two mirrors in a very similar style with closely related scenes have both been claimed as Vulci products: (1) Rome, Villa Giulia, inv. 24872, with Athena to the left and the inscriptions, *Menerva*, *Hercle*, *Heφleta*, the last regarded (DE SIMONE, *Entleh.* I, p. 69) as a variant of Heplenta: GERHARD, *ES*, pl. 341,2; FISCHER-GRAF, *Vulci*, pp. 93-4 no. V62, pl. 26,1; (2) Whereabouts unknown, formerly Viterbo, Bazzichelli collection, with Iolaos (Iule) to the left: KLÜGMANN–KÖRTE, *ES*, pl. 56; FISCHER-GRAF, *Vulci*, p. 93 no. V61. Both show Hippolyta also wearing *periamma* bands over her *chiton* with a medallion where they cross between her breasts and give a slightly different form to the broken Amazonian battle-axe below her. Although they show Herakles wearing his lionskin in the same way and the mirror formerly in Viterbo also has him gripping it in exactly the same fashion, they do not have him otherwise naked as here but wearing a breastplate with a *chiton* underneath it; also, slewed into a more vertical posture by the exigencies of the three-figure composition, he has his club raised instead of lowered. For quite different versions of the same scene on mirrors, see the Praenestine example (KLÜGMANN–KÖRTE, *ES*, pl. 57) and the Vulci mirror, Florence inv. 73645 (GERHARD, *ES*, pl. 136; FISCHER-GRAF, *Vulci*, p. 95 no. V68); a conversation group in Perugia includes Herakles (*Hercle*) and Hippolyta (*Heplenta*): KLÜGMANN–KÖRTE, *ES*, pl. 58; G. MATTEUCIG in *AJA* L, 1946, p. 63 fig. 2.

Engraving Characteristics. Where adequately preserved, the engraving seems to have been at least of moderate depth and in general even, firm and precise, the one notable lapse being the miscalculation in the layout of the lower palmette of the *obv.* extension decoration. The rather elongated torsos seem to have been Late Classical in character, as probably also the mannerism of the long, slender lionskin.

Fabric and Date. Etruscan, probably Vulci, Late Classical. The mirrors in Rome and formerly in Viterbo with scenes closely related to that in Cambridge (see above) have already been identified by FISCHER-GRAF (*Vulci*, pp. 93-4) as Late Classical Vulci products. Both these mirrors were assigned by G. CAMPOREALE (*StEtr* XXVII, 1959, p. 122) to a single artist who may here most conveniently be termed the Hippolyta Master; FISCHER-GRAF, (*Vulci* pp.

94-5 nos. V63-5) has added three more mirrors with quite different subjects, two of these last already linked together by J. G. SZILAGYI (cited *ibid.*), but has deleted the Herakles and Hippolyta mirror in Florence (see above) previously ascribed to the same artist by CAMPOREALE. Its preservation naturally makes any attribution of the Cambridge mirror difficult and uncertain. Nevertheless, it seems clear that its two-figure composition achieves a finer and more powerful scene than its crowded three-figure counterparts in Rome and once in Viterbo. These other two versions seem likely to be later debased adaptations of the Cambridge group and the sequence of degradation may be Cambridge – once Viterbo – Rome. The Cambridge mirror also shows remarkable correspondences with the other two both in composition and in detail, particularly on the figure of Hippolyta, and these may justify the tentative suggestion that it may be a finer, earlier work of this same Hippolyta Master.

26. Mirror. Figs. 26 a-d

Inv. CAM.440. Unprovenanced. Transferred 'from the Museum of Archaeology and Ethnology' at an unspecified date.

Unpublished.

Bronze. Surface heavily pitted. Completely and harshly cleaned to bare metal and coated with heavy seal. On technique, see Appendix.

D. 14.25 cm. H. 18.15 cm. H. of disc and extension 15.75 cm. W. at bottom of extension 1.55 cm. H. of tang 2.5 cm. Wt. 287.6 gm. Composition (see p. 54): copper 94%, tin 4.93%.

Circular mirror, its outline curving out below into a small tapered extension with, below it again, a bifurcated tang to secure it in missing handle.

Obv. Slightly convex reflecting face. Plain edge of slightly offset oblique profile. Top of extension and lower part of disc above engraved with pair of volutes with seven-petalled palmette above and, at each side, small hooked tendril and stem carrying heart-shaped ivy leaf. Engraving of volutes still of moderate depth, but that of palmette and ivy leaves worn so faint as to be now only just traceable.

Rev. Unengraved.

Fabric and Date. Etruscan, most probably Early Classical or beginning of Classical. For shape, cf. *CSE* Bologna I, no. 27; for engraved *obv.*, cf. loosely *op.cit.*, no. 41.

APPENDIX

THE COMPOSITION OF THE MIRRORS

P. T. Craddock and C. Buck
British Museum Research Laboratory

Technical. The samples were taken with a small drill by Dr J. A. Charles of the University Department of Metallurgy and Materials Science in Cambridge and Julie Dawson, Assistant Keeper in Charge of Conservation in the Antiquities Department at the Fitzwilliam Museum. Between five and thirteen milligrams of clean metallic turnings were analysed by atomic absorption spectrometry using the same apparatus and methods that were employed to analyse about 800 Etruscan bronzes, including about 200 mirrors principally from the British Museum (CRADDOCK, 1986). Full details of the methodology are given in HUGHES *et al.*, 1976.

The precision of the analyses is about +/- 1% for the major elements and +/- 20% for trace elements. All the listed elements could be detected down to about 0.01% in the metal.

Discussion. The mirrors were all of bronze with relatively high tin contents and low lead contents, entirely typical of Etruscan mirrors (CRADDOCK, 1986). Only one mirror here, the beaten mirror, **14**, has what may be considered a significant lead content (4.87%) and this piece is of dubious authenticity, although Late Etruscan mirrors can sometimes have several percent of lead in the alloy (see histograms).

The composition of Etruscan mirrors is markedly different from that of other contemporary cast Etruscan metalwork. This tends to have less tin and a much greater prevalence of leaded alloys, although of rather variable composition. The mirror alloys are much more like the sheet metal used for vessels and armour. Similar trends are observable in contemporary Greek bronzes, although the Etruscan mirrors do seem to have been of a more consistent and superior alloy than their Greek counterparts (CRADDOCK, 1977).

The only detailed metallographic study of the mirrors was carried out many years ago by Panseri and Leoni (PANSERI and LEONI, 1957). They considered the mirrors had been cast more or less to shape in a stone mould, then, prior to hammering, heated at about 600°C. This was necessary to homogenize the metal and change the brittle and hard $\alpha + \delta$ eutectoid phase to the more ductile α phase. If there is more than 10% of tin in the copper then there is a strong chance of the eutectoid being present. Its absence in the ten specimens examined by Panseri and Leoni suggest that it had been removed from the mirrors by heat treatment. The metallographic structure of the mirrors showed they had then been extensively cold-worked by hammering and then annealed.

The mirrors in Cambridge are very likely to have been made in this way as they are typical examples of Etruscan mirrors in both form and composition.

Analyses of the Mirrors

No	Cu	Sn	Pb	Ag	Fe	Ni	Zn	As	Sb	Au	Bi	Co	Mn	Cd
1	93.5	4.15	0.944	0.102	0.056	0.054		0.20	0.07		0.181	0.016	0.0013	
2	89.0	10.51	0.165	0.030	0.200	0.037	0.010	0.28	0.08			0.023		
3	No analysis													
4	85.5	8.97	0.065	0.014	0.087	0.034		0.11	0.04			0.119		
5	86.5	12.32	0.191	0.033	0.167	0.041		0.07	0.05			0.059		
6	88.5	10.54	0.114	0.044	0.370	0.032		0.20	0.06					
8	87.5	10.59	0.088	0.026	0.042	0.023		0.13	0.04			0.009	0.0022	
9	86.0	12.68	0.101	0.023	0.086	0.028		0.19	0.05			0.019	0.0009	
10	85.5	12.99	1.107	0.055	0.057	0.036		0.10	0.09			0.015	0.0020	
11	91.0	7.83	0.144	0.037	0.088	0.027		0.07	0.06			0.010	0.0009	
12	89.5	9.66	0.031	0.011	0.287	0.036		0.19	0.05			0.036		
13	No analysis													
14	85.0	9.30	4.869	0.054	0.098	0.033	0.010	0.17	0.06		0.063	0.036		0.0038
15	89.5	9.34	0.021	0.023	0.096	0.016		0.20	0.04				0.0049	
16	89.5	6.41	1.350	0.034	0.177	0.043		0.17	0.04			0.036		
17	86.5	12.02	0.025	0.017	0.045	0.043		0.19	0.04			0.076	0.0039	
18	86.5	10.59	0.036	0.019	0.065	0.036		0.27	0.05			0.097	0.0007	
19	85.0	13.51	0.167	0.031	0.158	0.033	0.040	0.24	0.05			0.031	0.0007	
20	83.0	12.65	0.131	0.030	0.473	0.035	0.020	0.51	0.07	0.014		0.030	0.0030	
21	86.0	12.73	0.083	0.032	0.085	0.020		0.14	0.06			0.024	0.0014	
23	87.0	10.31	0.041	0.046	0.024	0.035		0.19	0.06	0.010				
24	87.5	10.55	0.020	0.030	0.085	0.040		0.07	0.05	0.010		0.032	0.0007	
25	91.5	6.44	0.318	0.041	0.101	0.035			0.08				0.0033	

Note that lead samples were provided only from **3** and **13** and that there were no metal samples at all from **7** (of bone) and **22** (unavailable for analysis). The identification serials for these analyses are SPQR 3292 - 3315 inclusive.

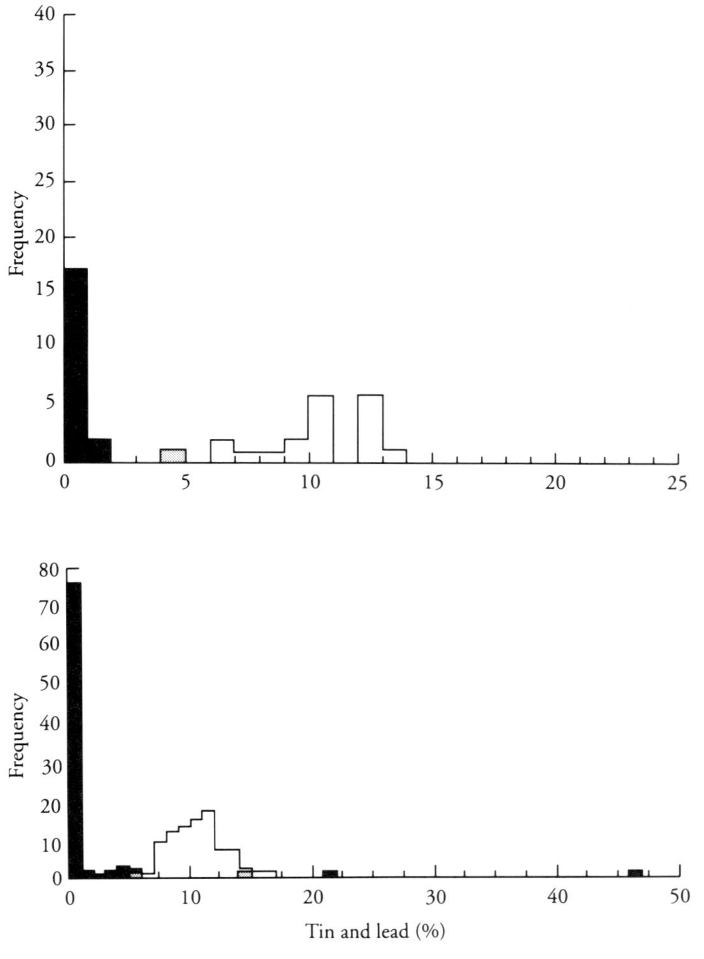

Histograms of lead and tin content of Etruscan Mirrors.
Above: Cambridge Collections. *Below:* The British Museum, the Ashmolean Museum and Museums in Copenhagen, after Craddock 1986 (p. 229, fig. 16).
Key: black= lead, white = tin, stippled = tin and lead coinciding.

References cited:

CRADDOCK, 1977 — P. T. CRADDOCK, 'The Composition of the Copper Alloys used by the Greek, Etruscan and Roman Civilisations, 2, The Archaic, Classical and Hellenistic Greeks', *Journal of Archaeological Science* IV, 1977, pp. 103-24.

CRADDOCK, 1986 — P. T. CRADDOCK, 'The Metallurgy and Composition of Etruscan Bronze', *StEtr* LII, 1986, pp. 211-71.

HUGHES *et al.*, 1976 — M. J. HUGHES, M. R. COWELL and P. T. CRADDOCK, 'Atomic Absorption Techniques in Archaeology'. *Archaeometry*, XVIII, 1976, pp. 19-36.

PANSERI and LEONI, 1957 — C. PANSERI and M. LEONI, 'The Manufacturing Technique of Etruscan Mirrors', *Studies in Conservation* III, 1957, pp. 49-63.

British Museum Research Laboratory File No. 5275.

INDEX

1. IDENTIFICATION SERIALS

(a) *Inventory numbers*

CAM-Prefix (Museum of Classical Archaeology)

CAM.343	**25**
CAM.440	**26**

E-Prefix (Fitzwilliam Museum)

E.101.1937	**7**

GR-Prefix (Fitzwilliam Museum)

GR.12.1864	**8**
GR.12*a*.1864	**9**
GR.19.1904	**10**
GR.1.1941	**13**
GR.10.1972	**11**
GR.2.1976	**12**
GR.3.1977	**14**

RC-Prefix (Museum of Archaeology and Anthropology)

RC.23.249	**16**
RC.23.317	**17** (= Z.23214*a*)

Z-Prefix (Museum of Archaeology and Anthropology)

Z.23214*a*	**17** (= RC.23.317)
Z.23214*b*	**18** (= Z.39534)
Z.23214*c*	**19**
Z.23214*d*	**20**
Z.23214*e*	**21**
Z.23214*f*	**22**
Z.23215*a*	**23**
Z.23215*b*	**24**
Z.39534	**18** (= Z.23214*b*)

No Prefix (Museum of Archaeology and Anthropology)

22.1332	**15**

Uninventoried (Corpus Christi College)

1 - **6** inclusive

(b) Loan numbers

Ant-Prefix (Fitzwilliam Museum), Corpus Christi College loan

Ant.199/103.35	**1**
Ant.199/103.36	**6**
Ant.199/103.37	**5**
Ant.199/103.38	**2**
Ant.199/103.39	**4**
Ant.199/103.43	**3**

2. CHRONOLOGY

Archaic – Early Classical	**15**
Early Classical	**1**, **7**
Early Classical – start of Classical	**23**, **26**
Late Classical	**2**, **8**, **25**; end:**6**
End of Late Classical – Hellenistic	**16**
Early Hellenistic	**11**, **12**
Early Hellenistic – start of Hellenistic	**4**
Hellenistic	**3**, **5**, **9**, **10**, **17**, **18**, **19**, **20**, **21**, **22**, **24**
Uncertain Authenticity	**13**, **14**

3 TYPOLOGY

(a) Discs

Circular	**1**, **8**, **9**, **15**, **20**, **23**, **25**, **26**;
with outer band:	**19**
with outer band, raised border	**10**, **17**, **24**
with raised border only	**4**, **18**
almost circular	**2** (with tendency to pear-shaped), **11** (with tendency to ovoid)
missing discs probably originally circular	**21**, **22**
Circular *appliqué*	**13**
Circular, lidded	**3**, **16** (lid lost)
Pear-shaped	**6**, **12**
'debased' pear-shaped	**5**, **14**

(b) Extensions

Tang simply proud of handle	**1**
Tapering with slightly curved sides, obtuse points	**11**, **23**, **25**, **26**
Concave-sided with squarish points	**2**, **8**
Markedly concave-sided, flaring below at angular points	**4**, **6**, **9**, **10**, **12**, **17**, **18**, **21**, **22**;
probably originally of this form (now incomplete)	**19**, **20**, **24**
Vestigial, with small angular points	**5**

8. GENERAL INDEX

ILLUSTRATIONS

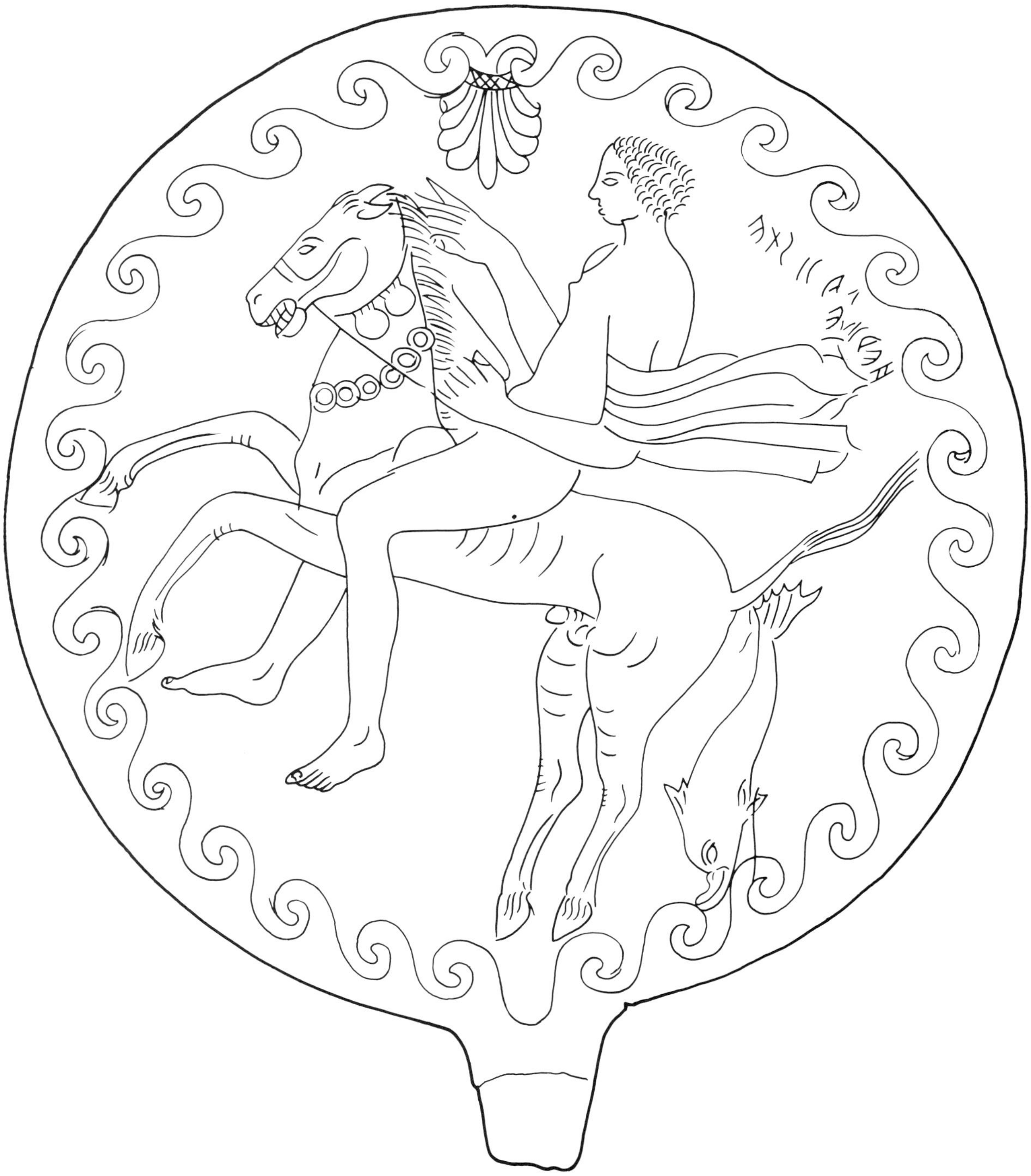

1a

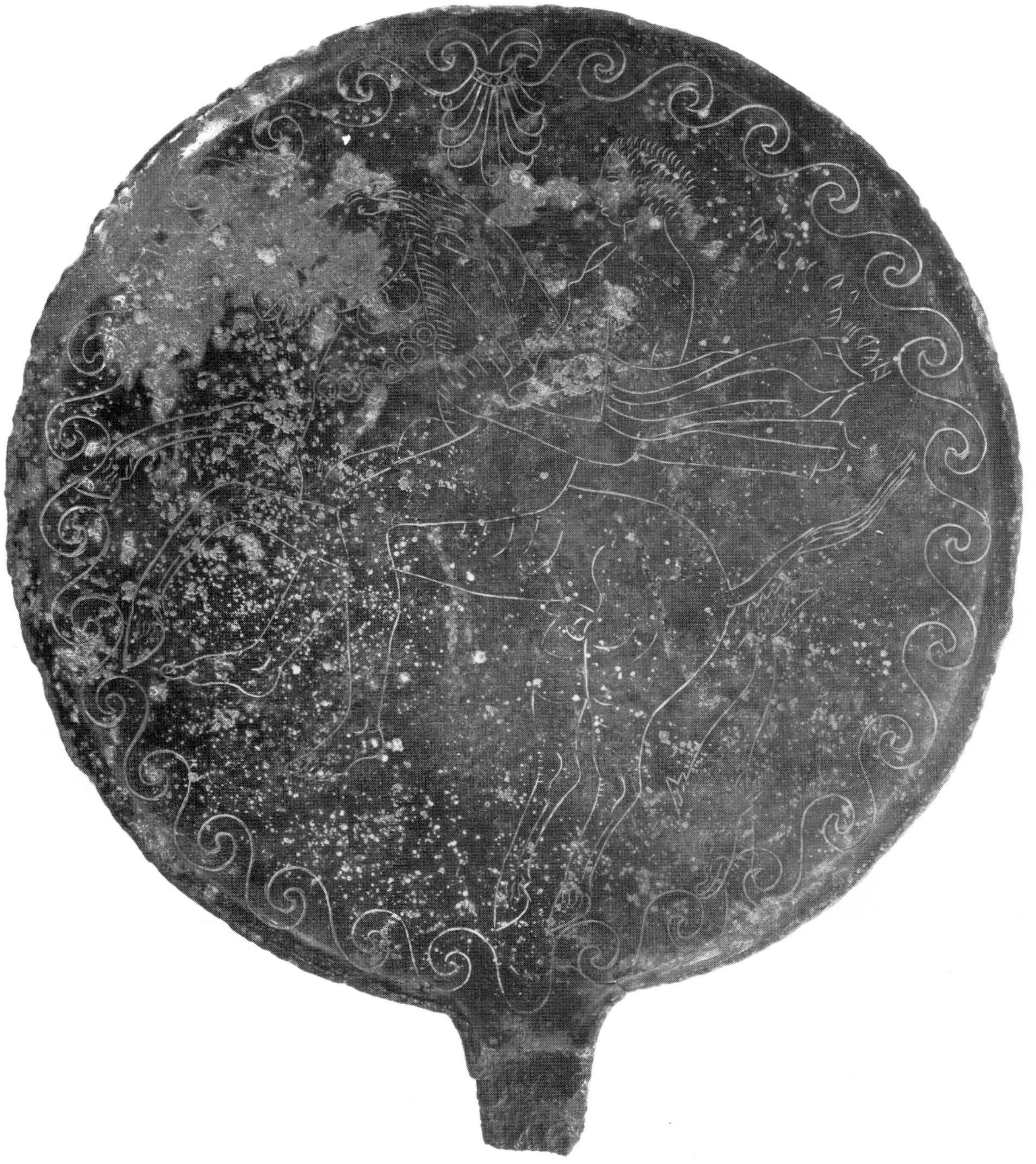

1b

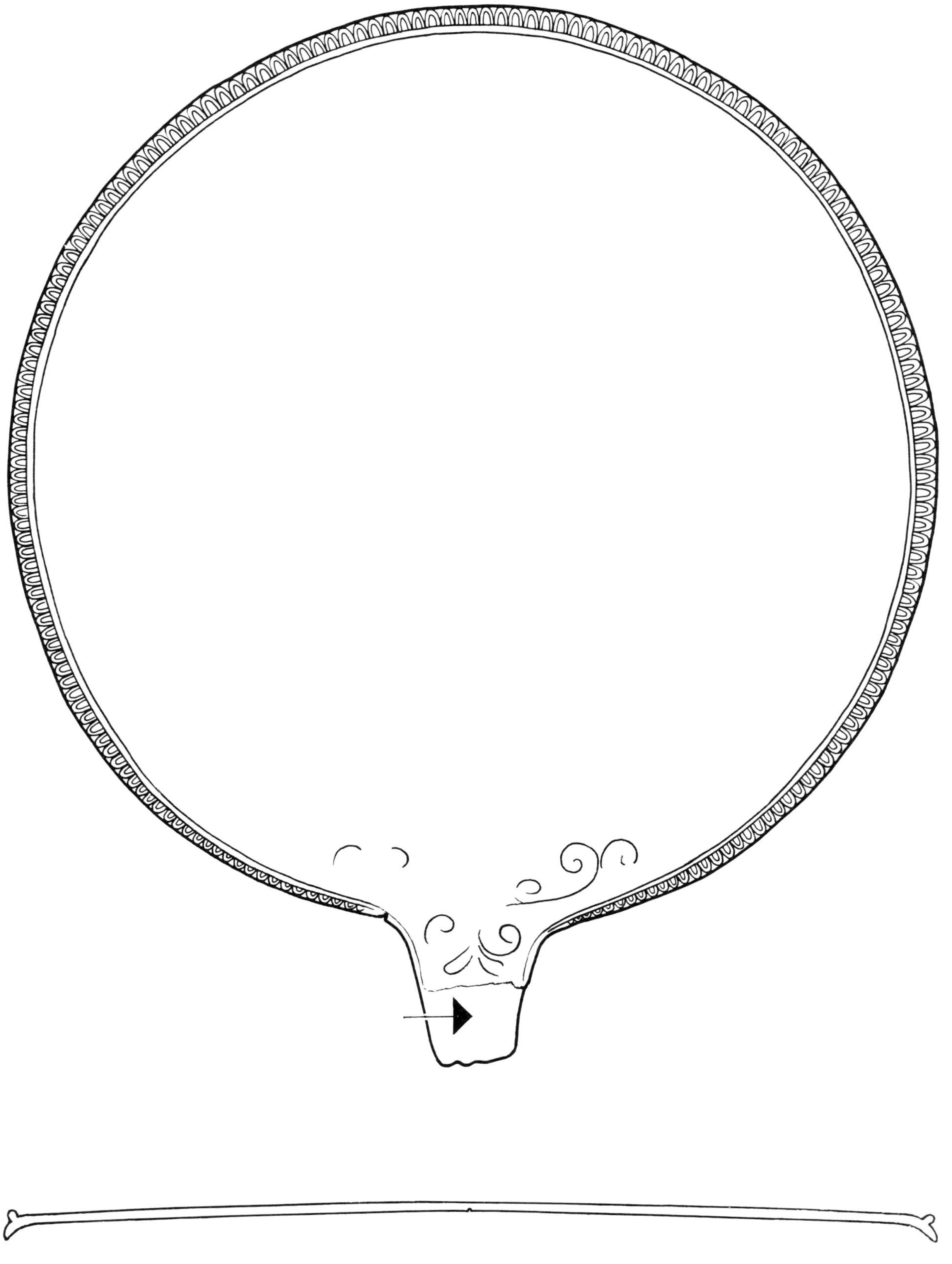

1c

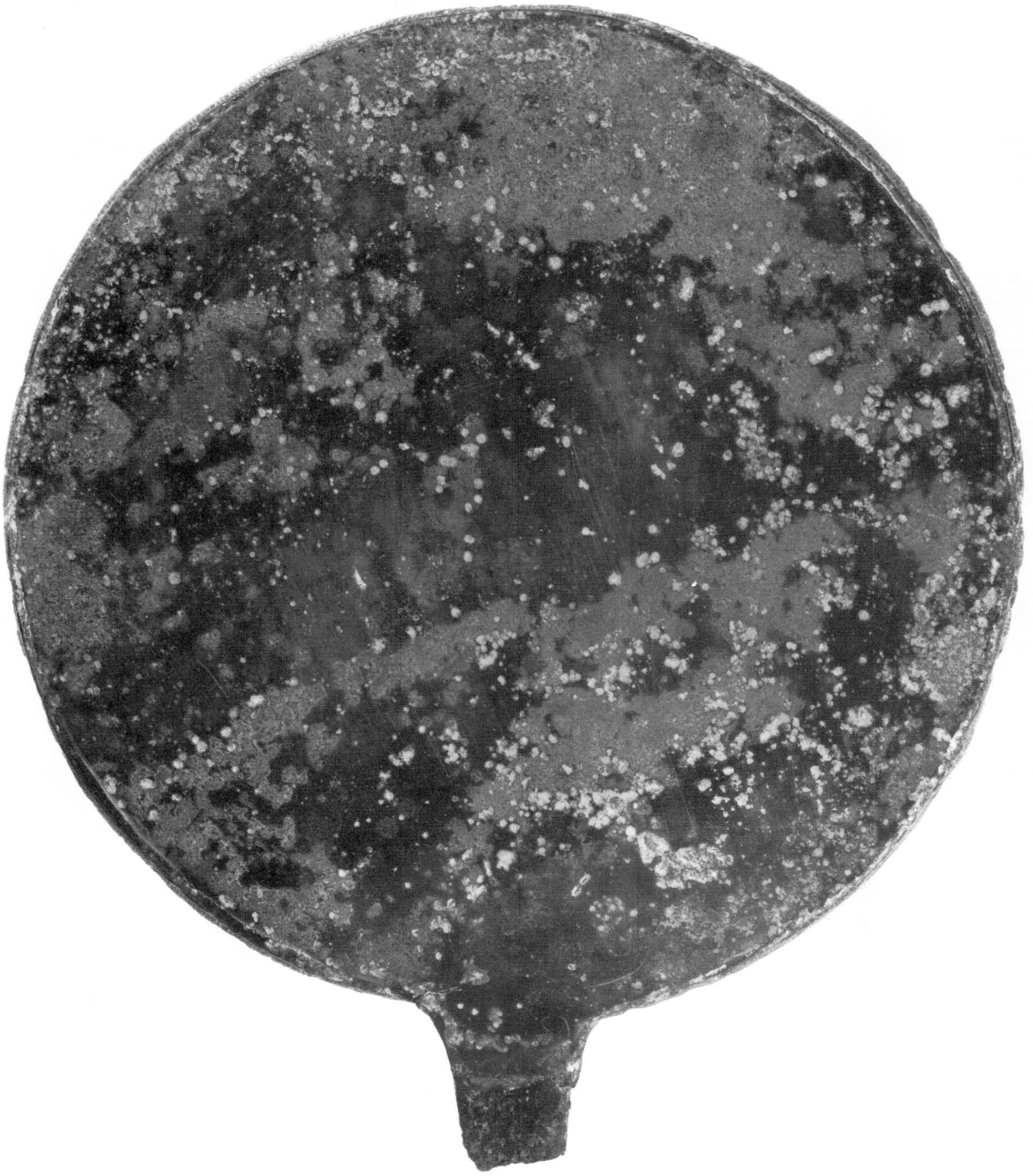

1d

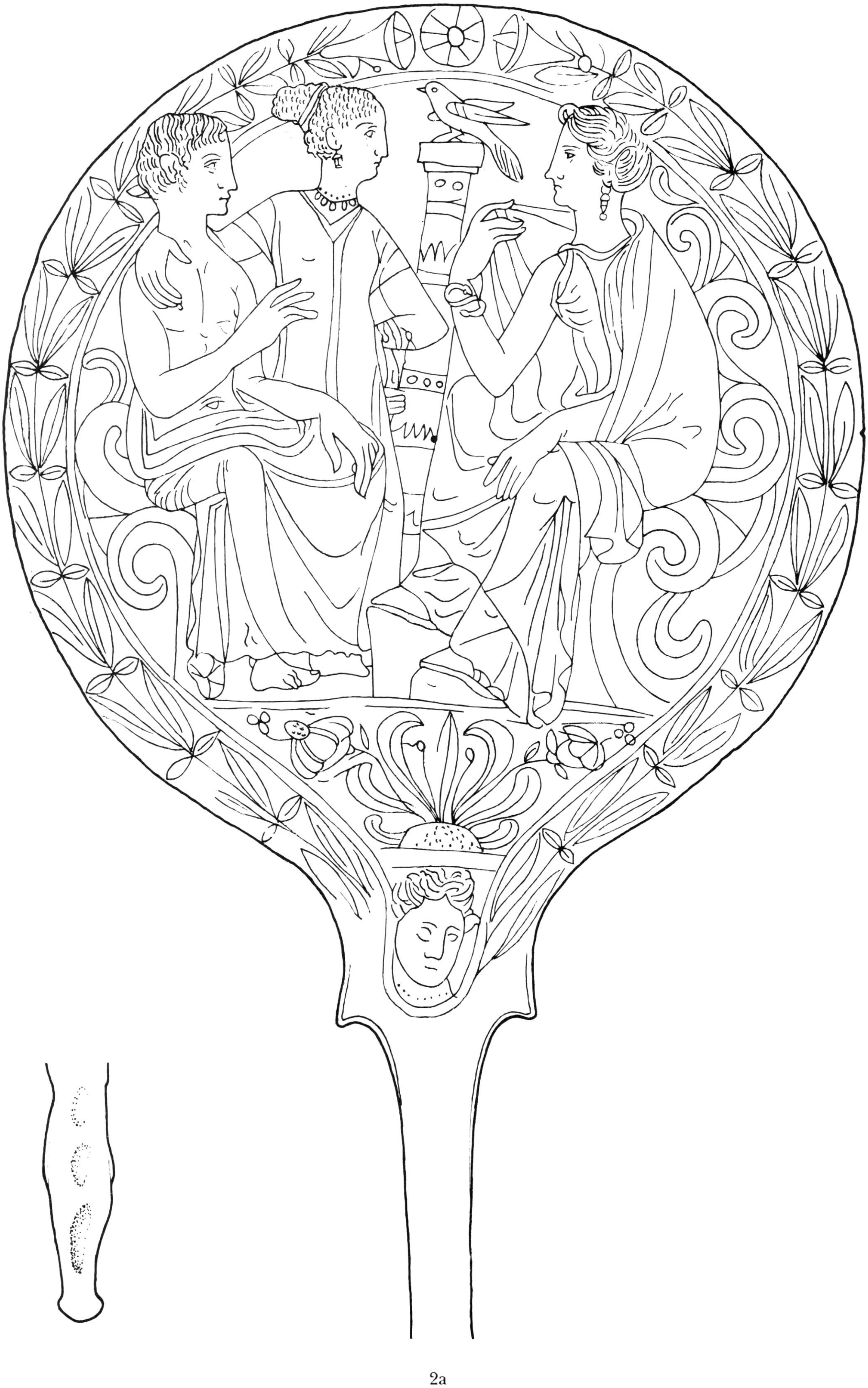

2a

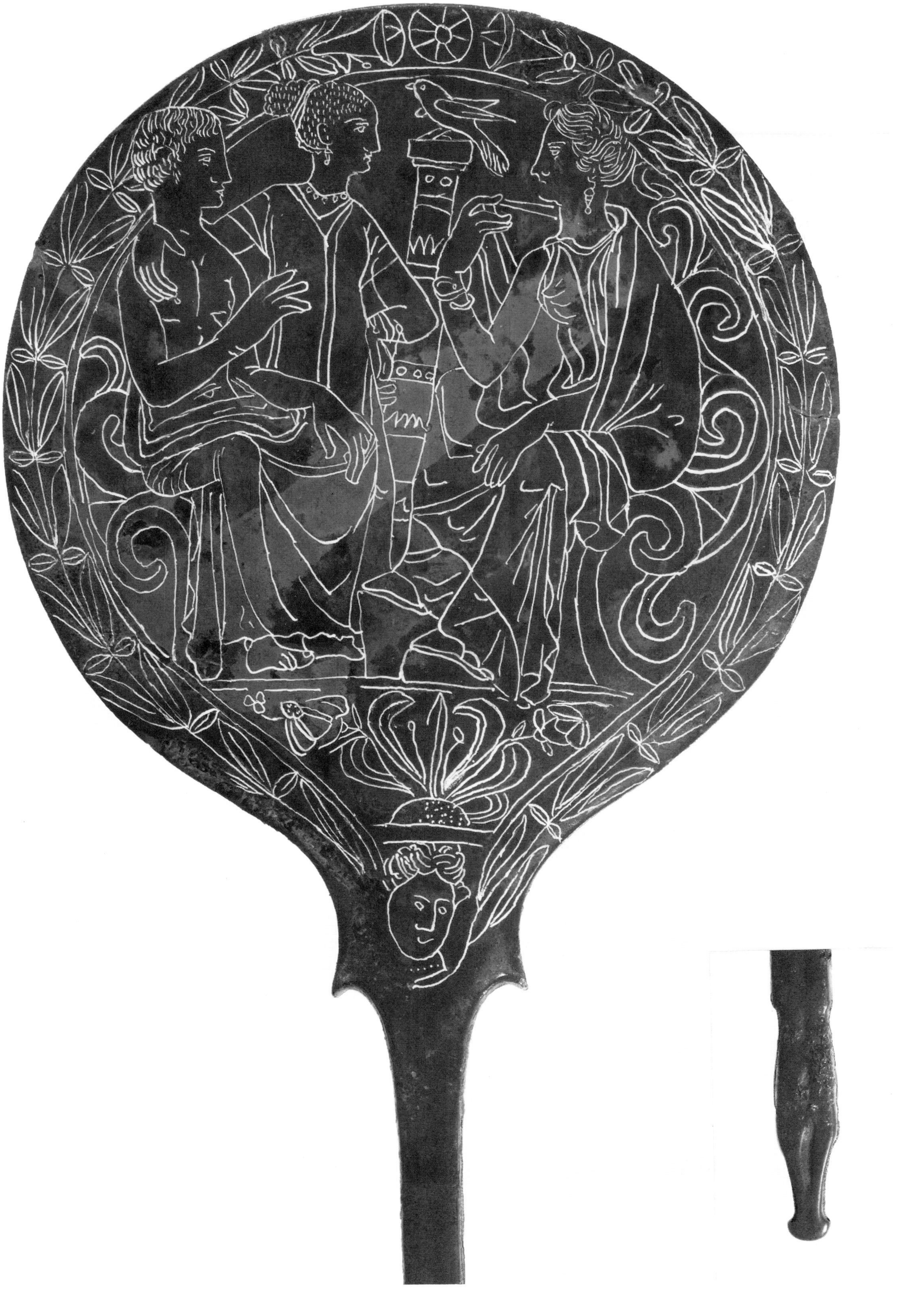

2b

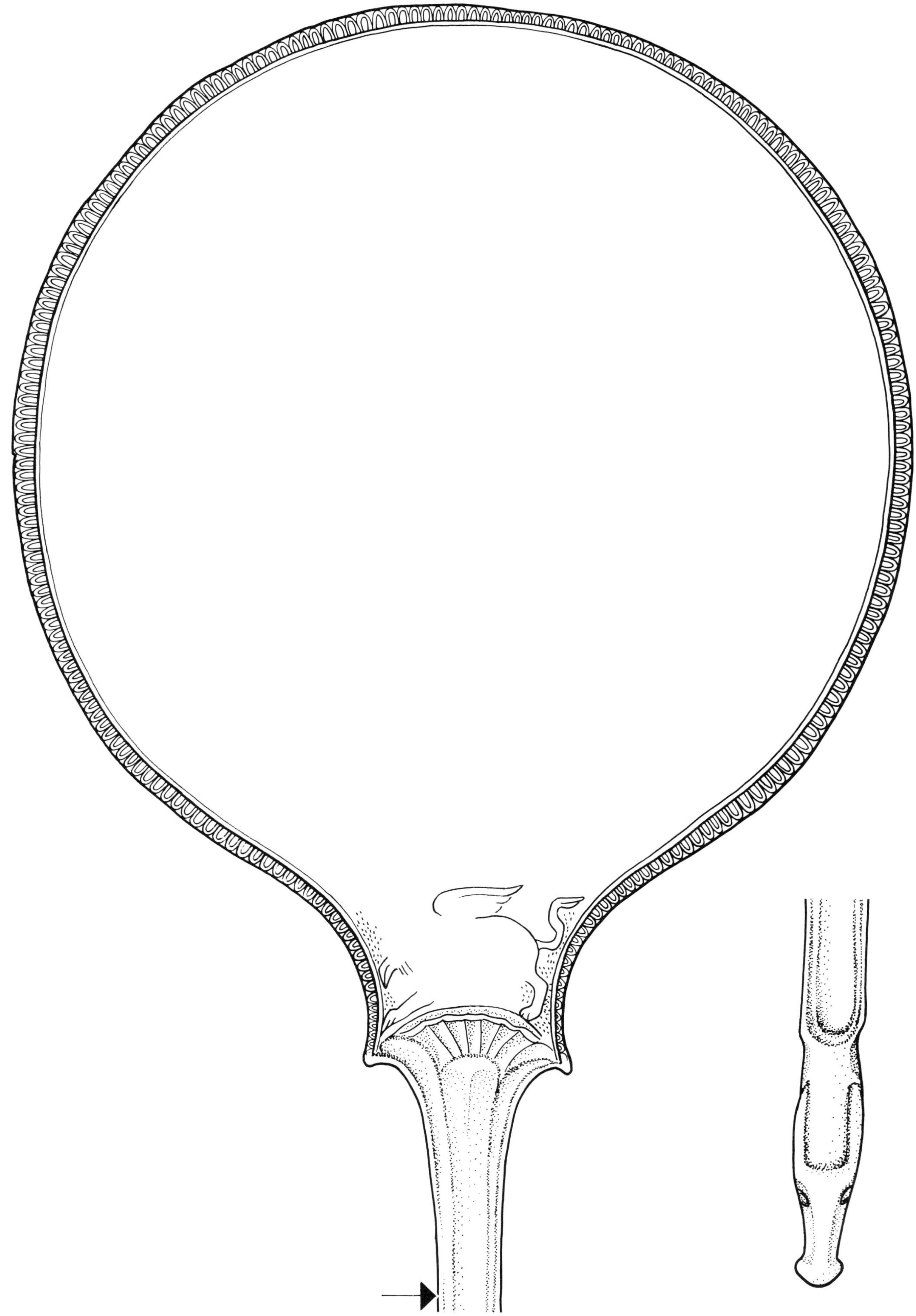

2c

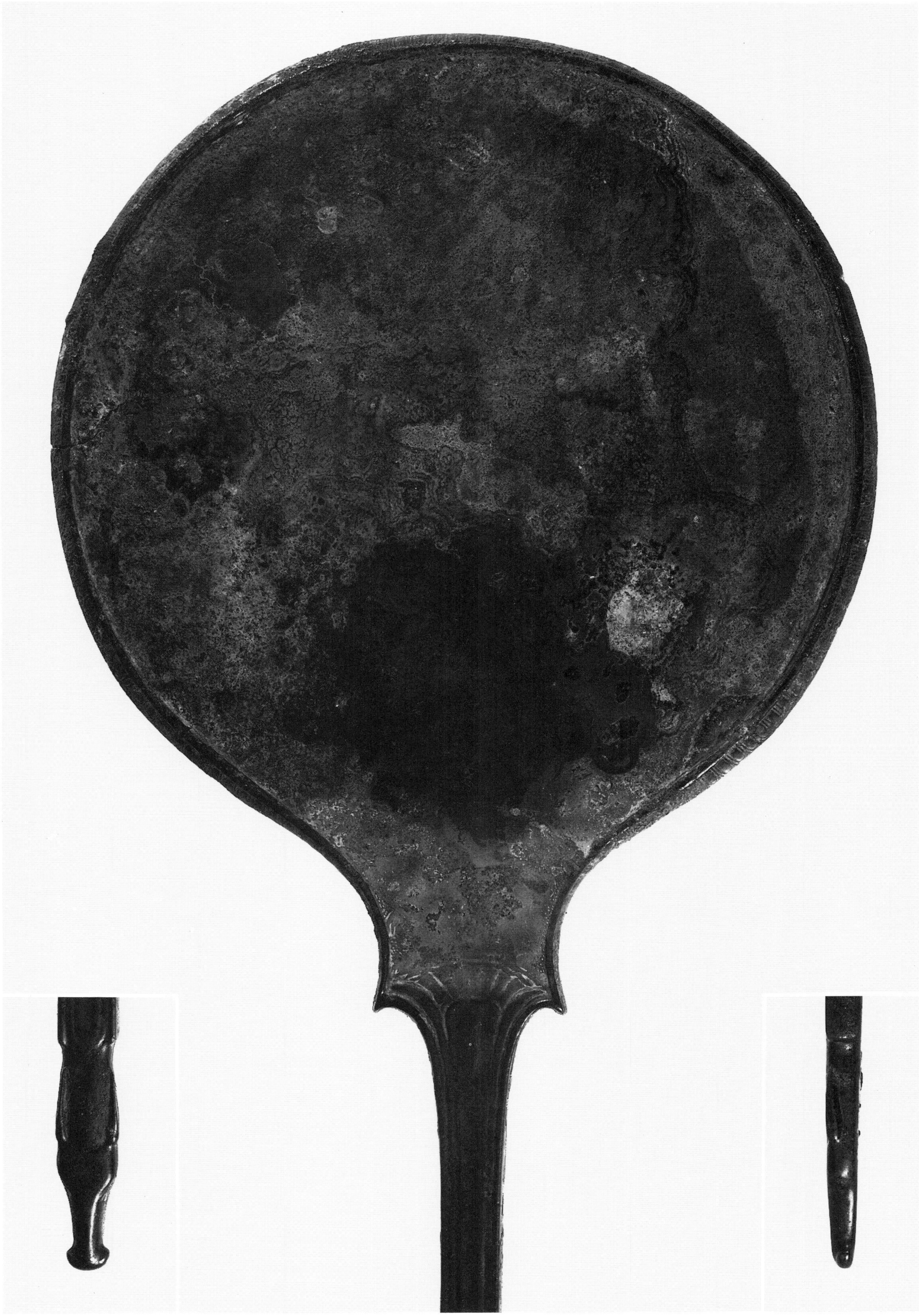

2d

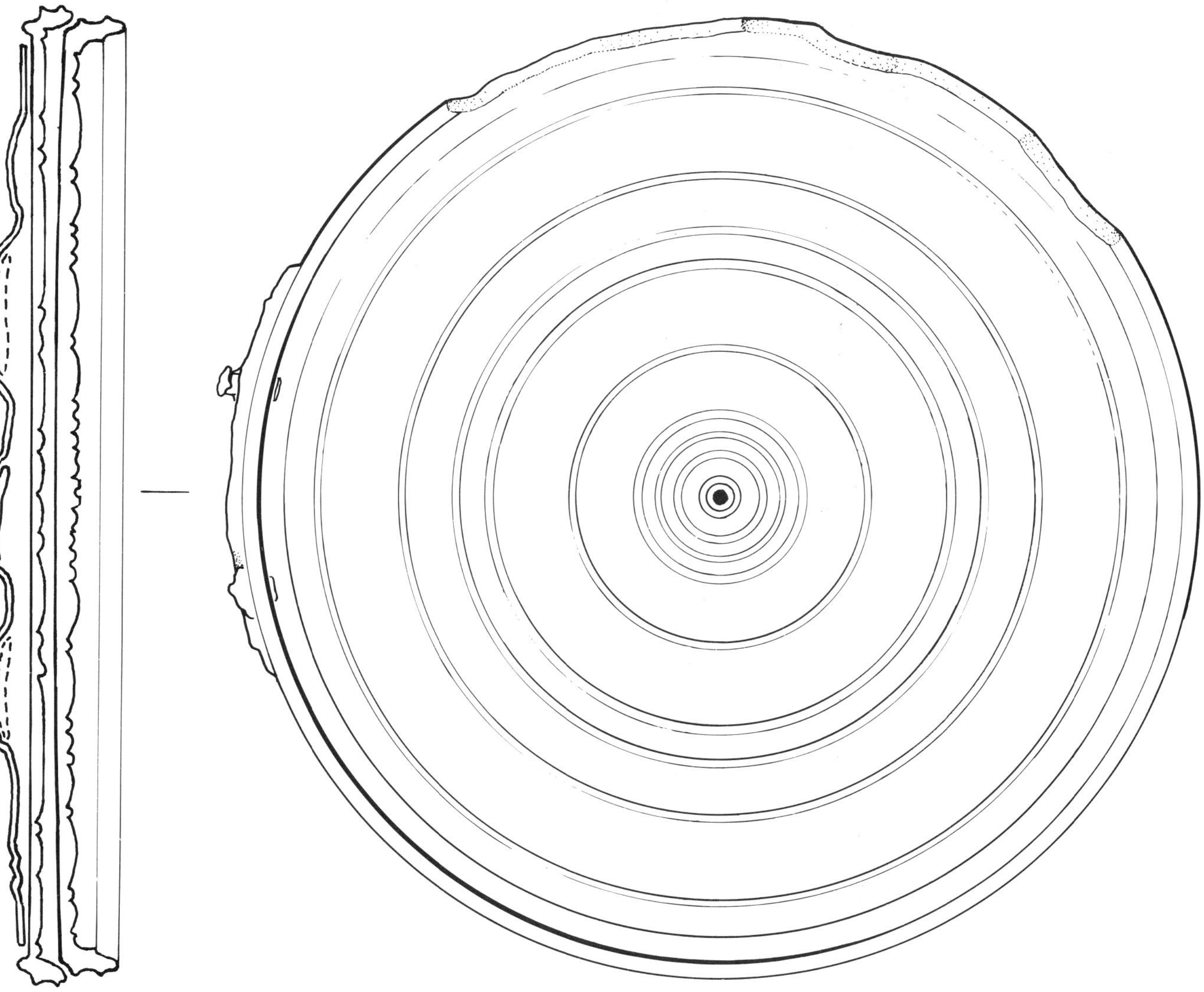

3c

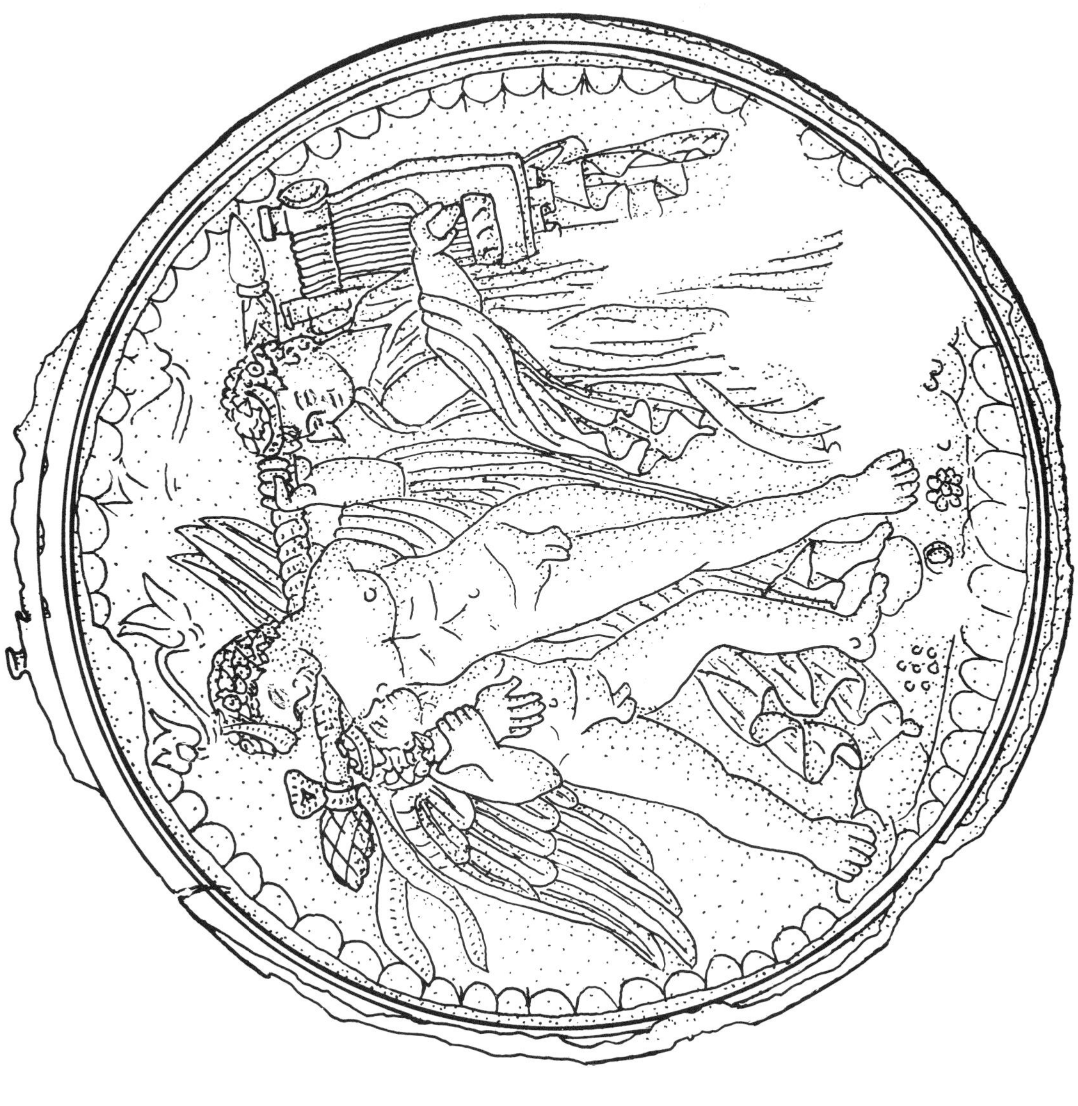

3a

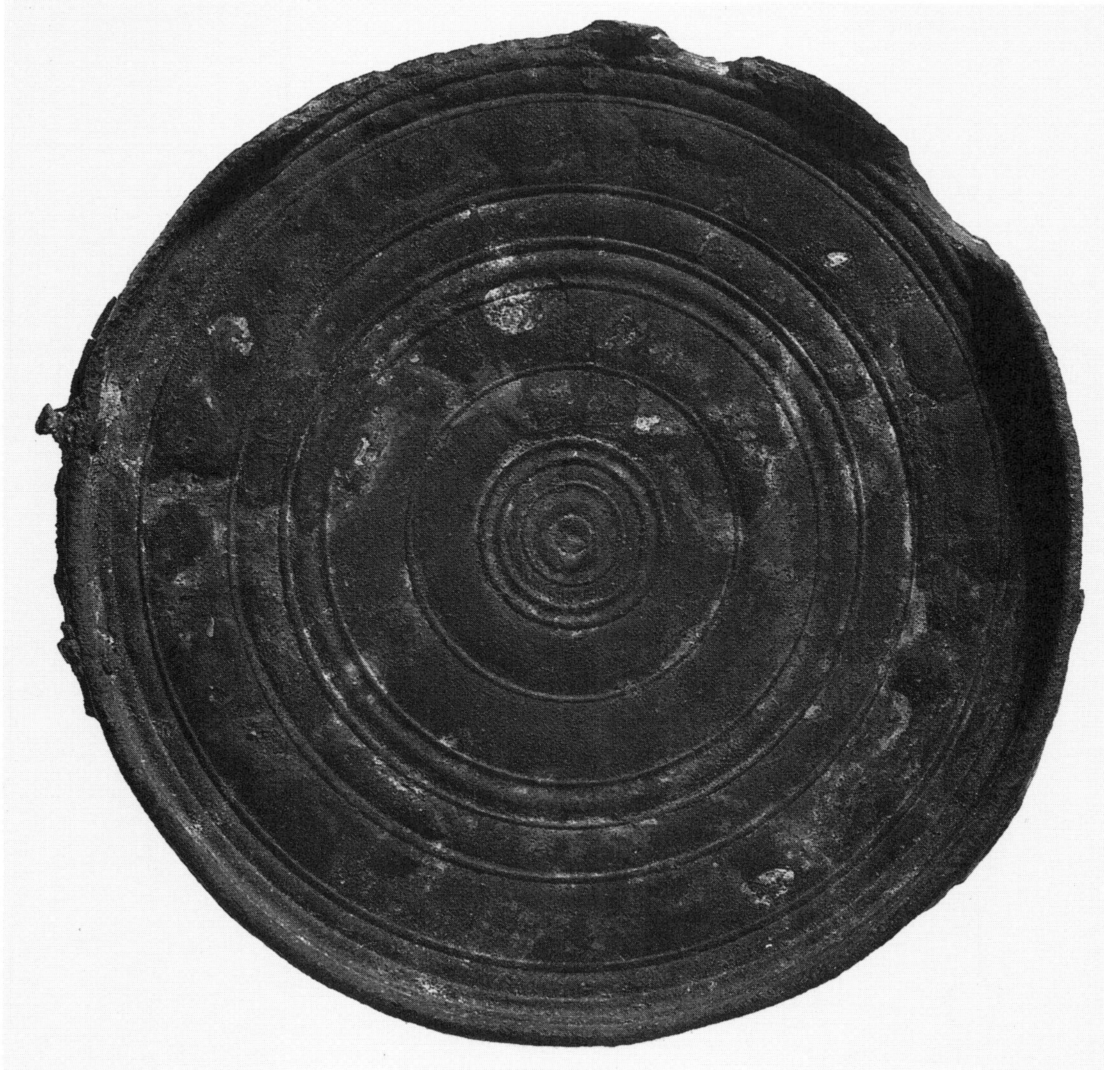

3d

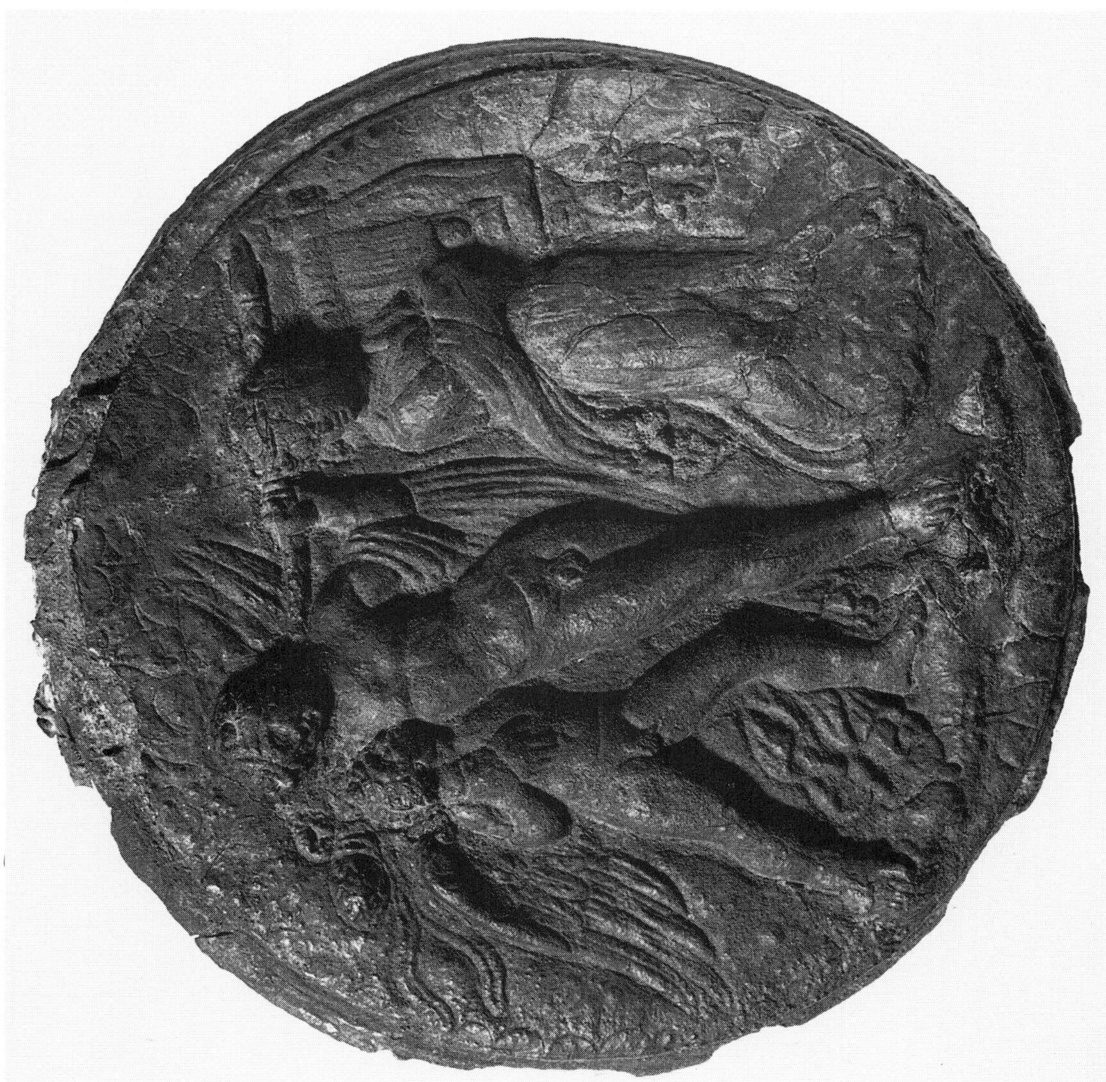

3b

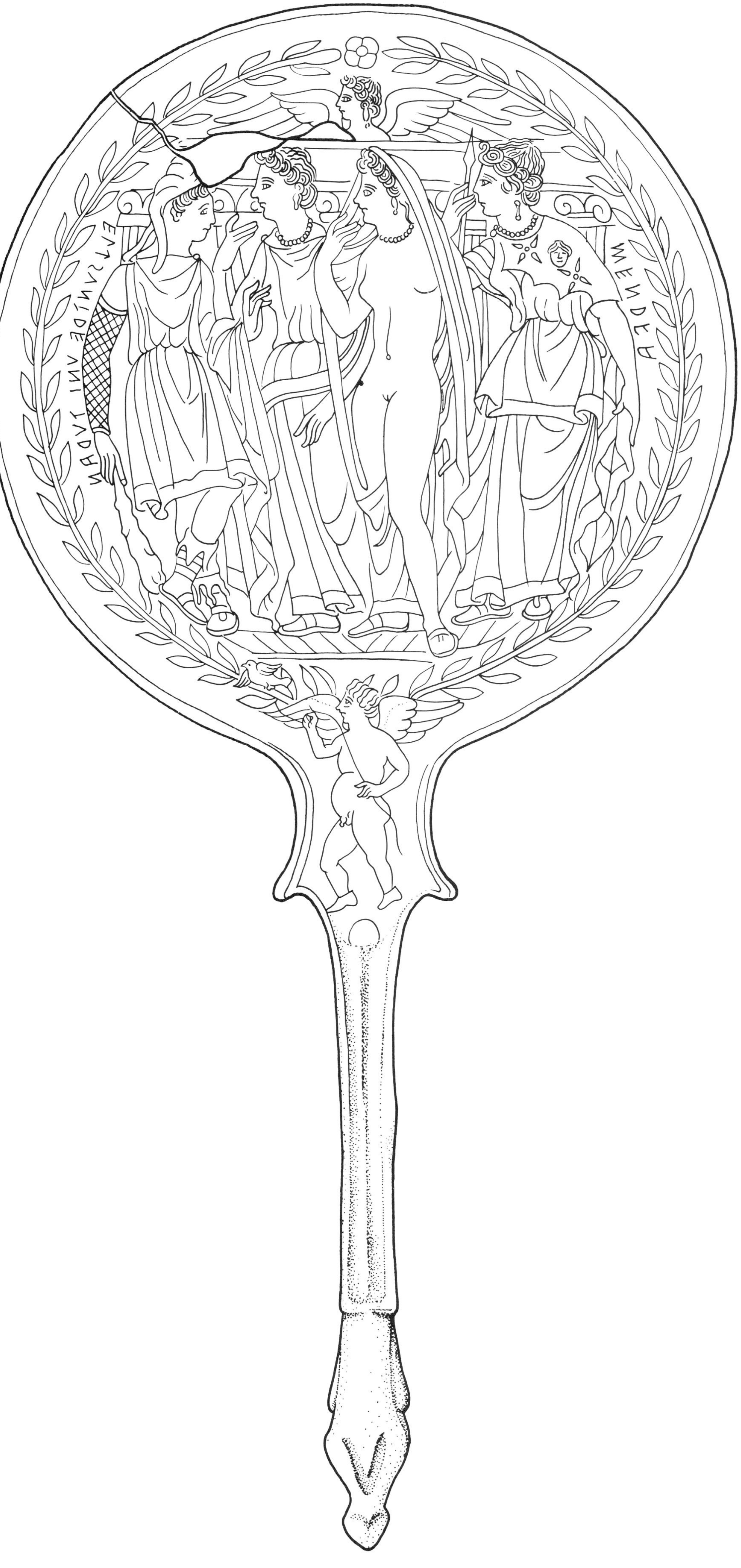

4a

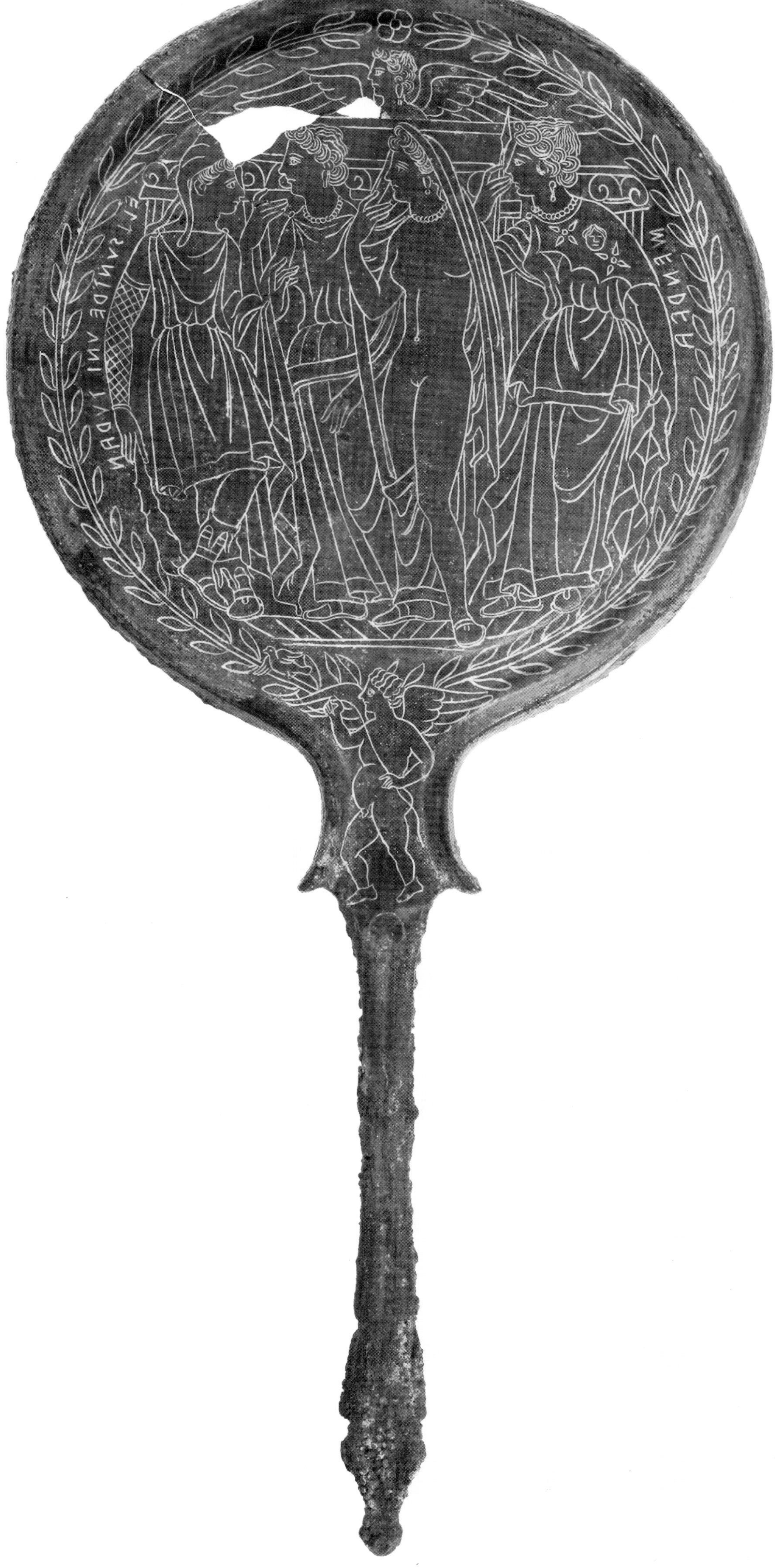

4b

CSE,
GREAT
BRITAIN
2, 4

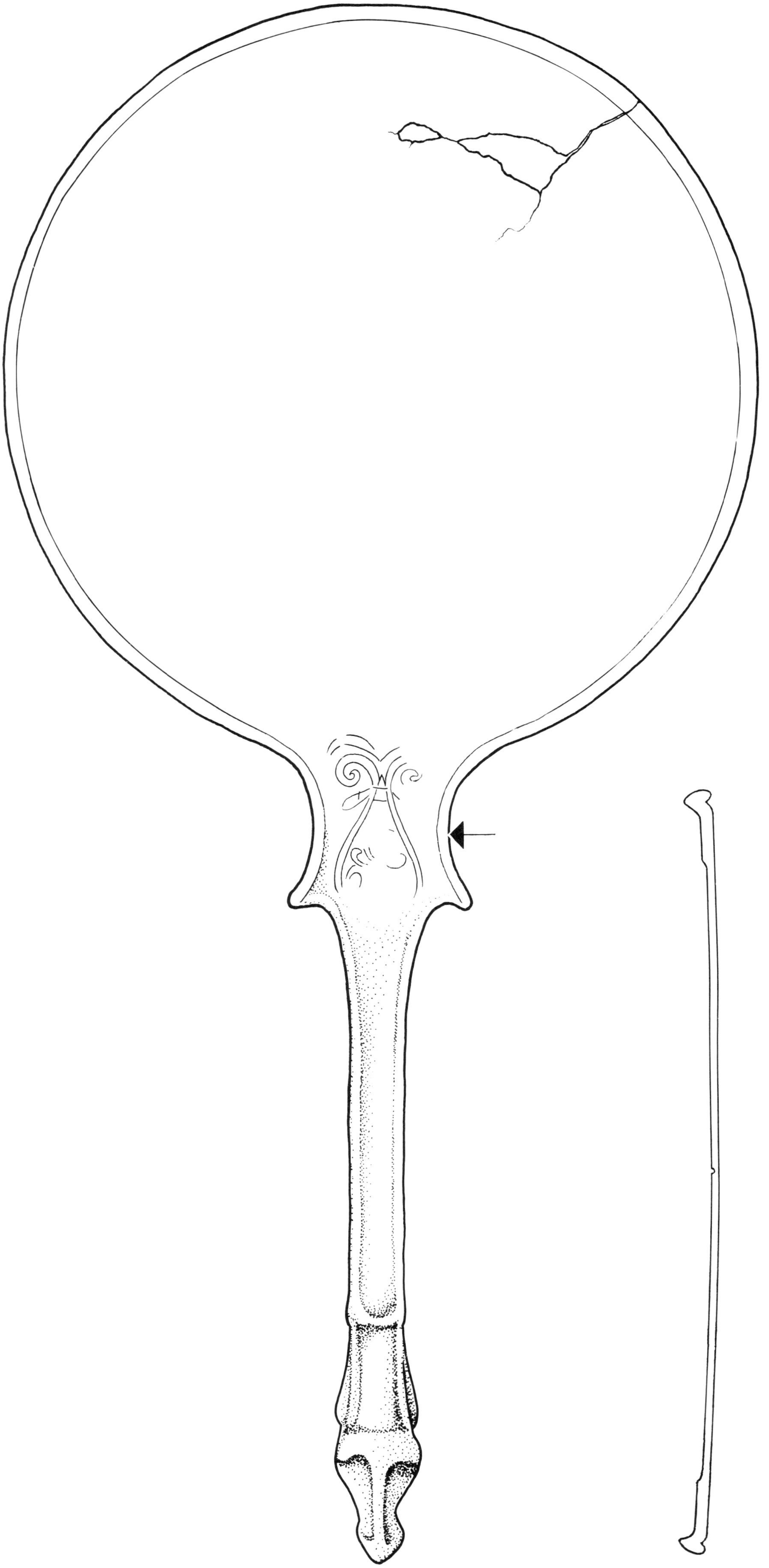

4c

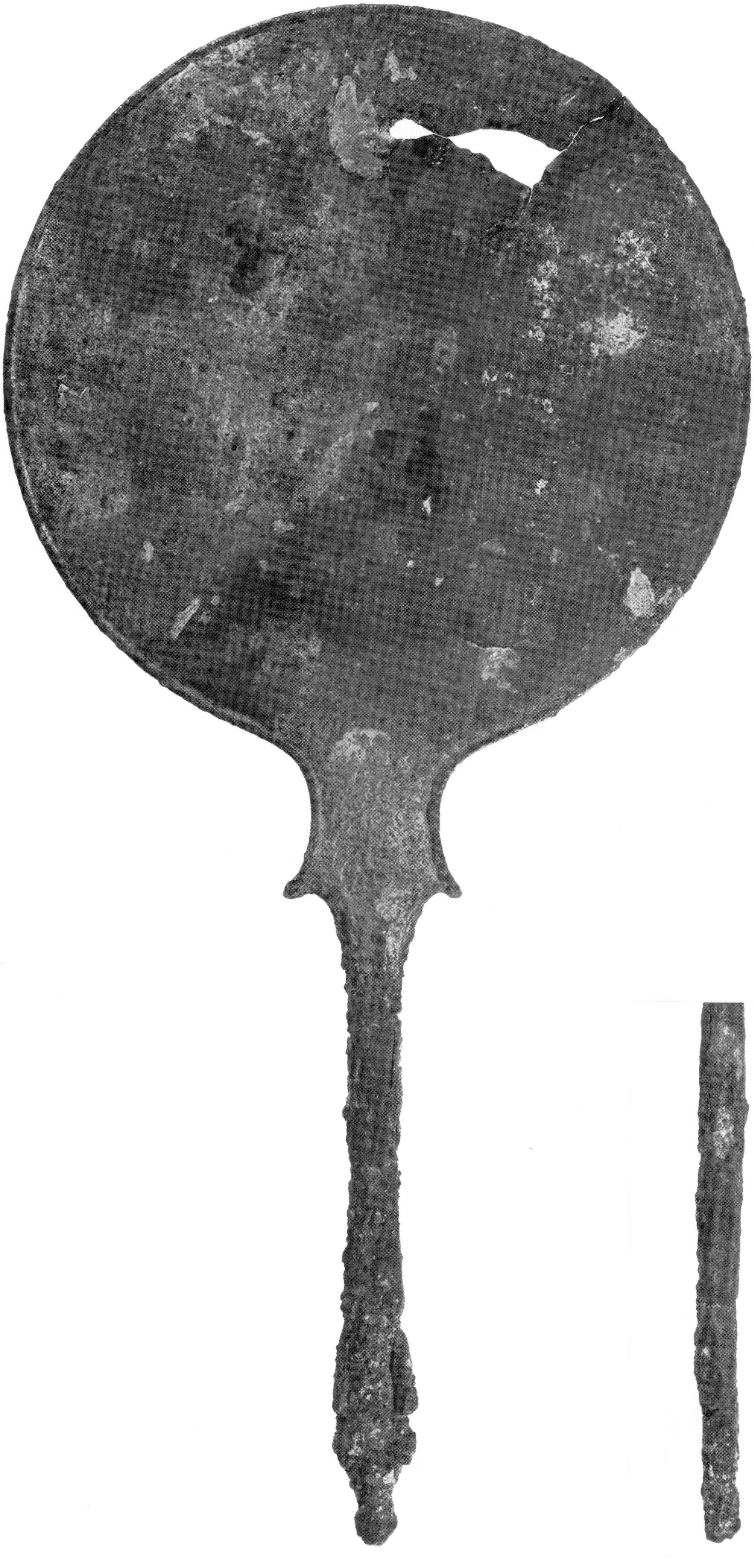

4d

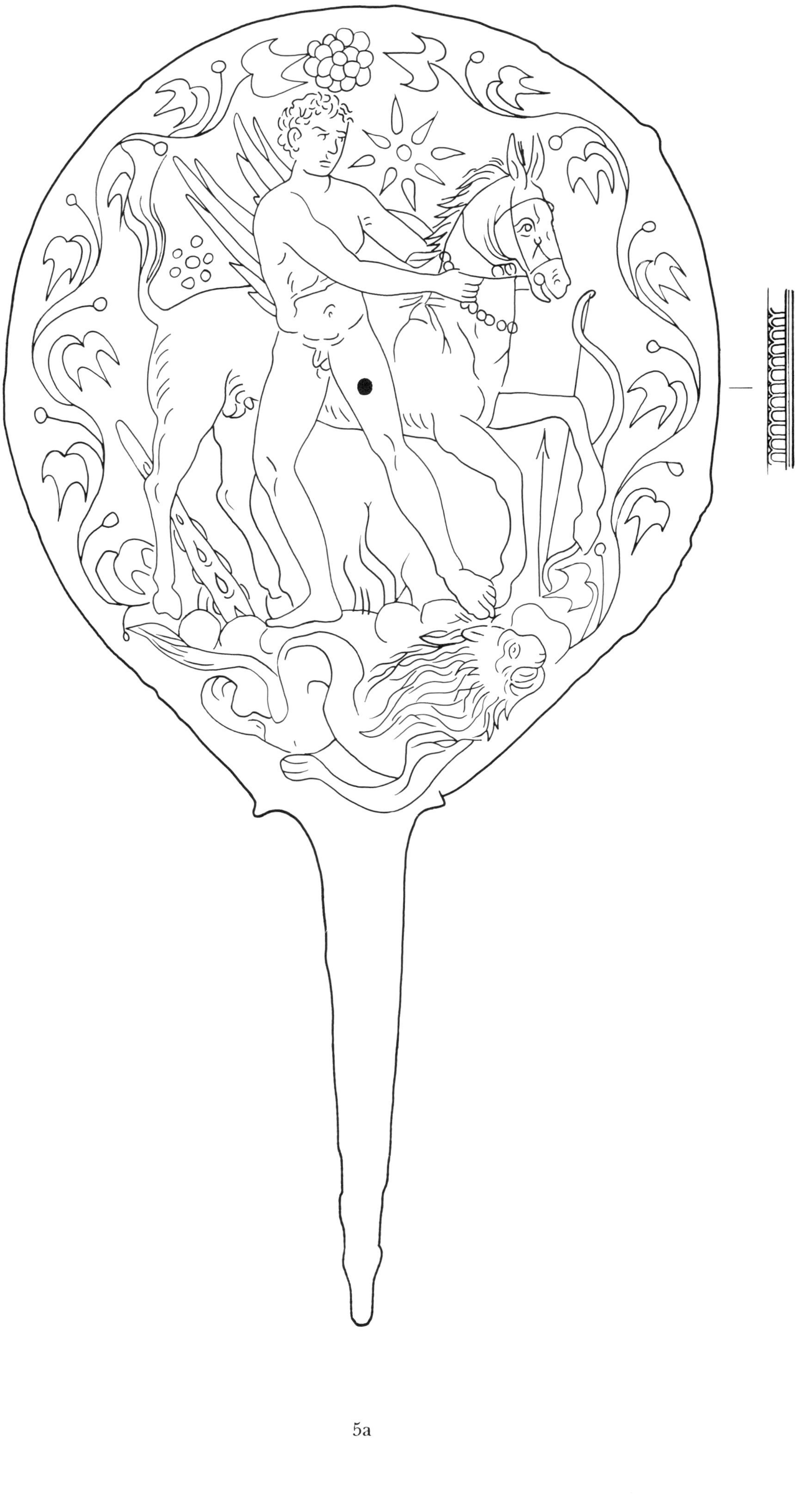

5a

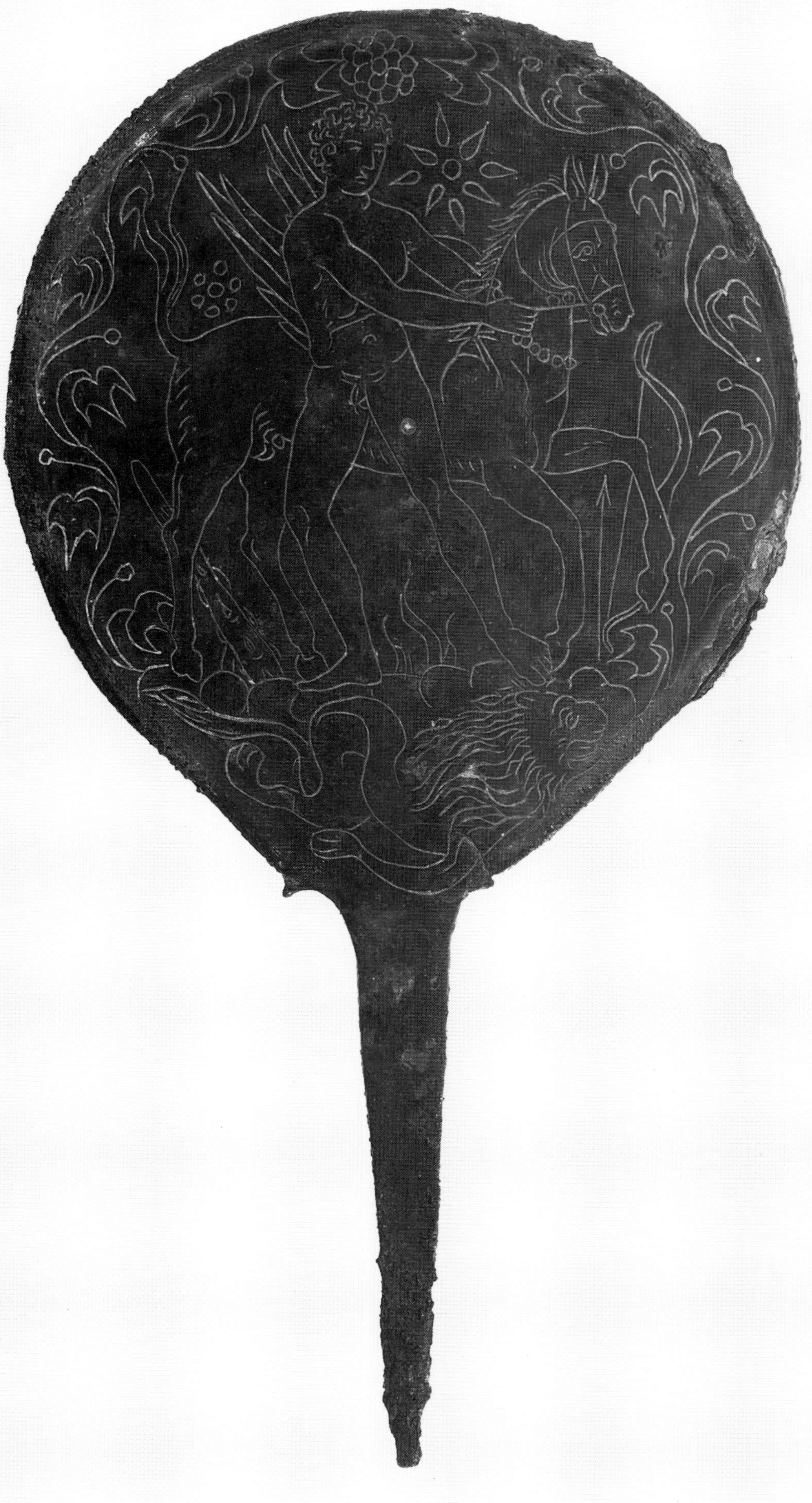

5b

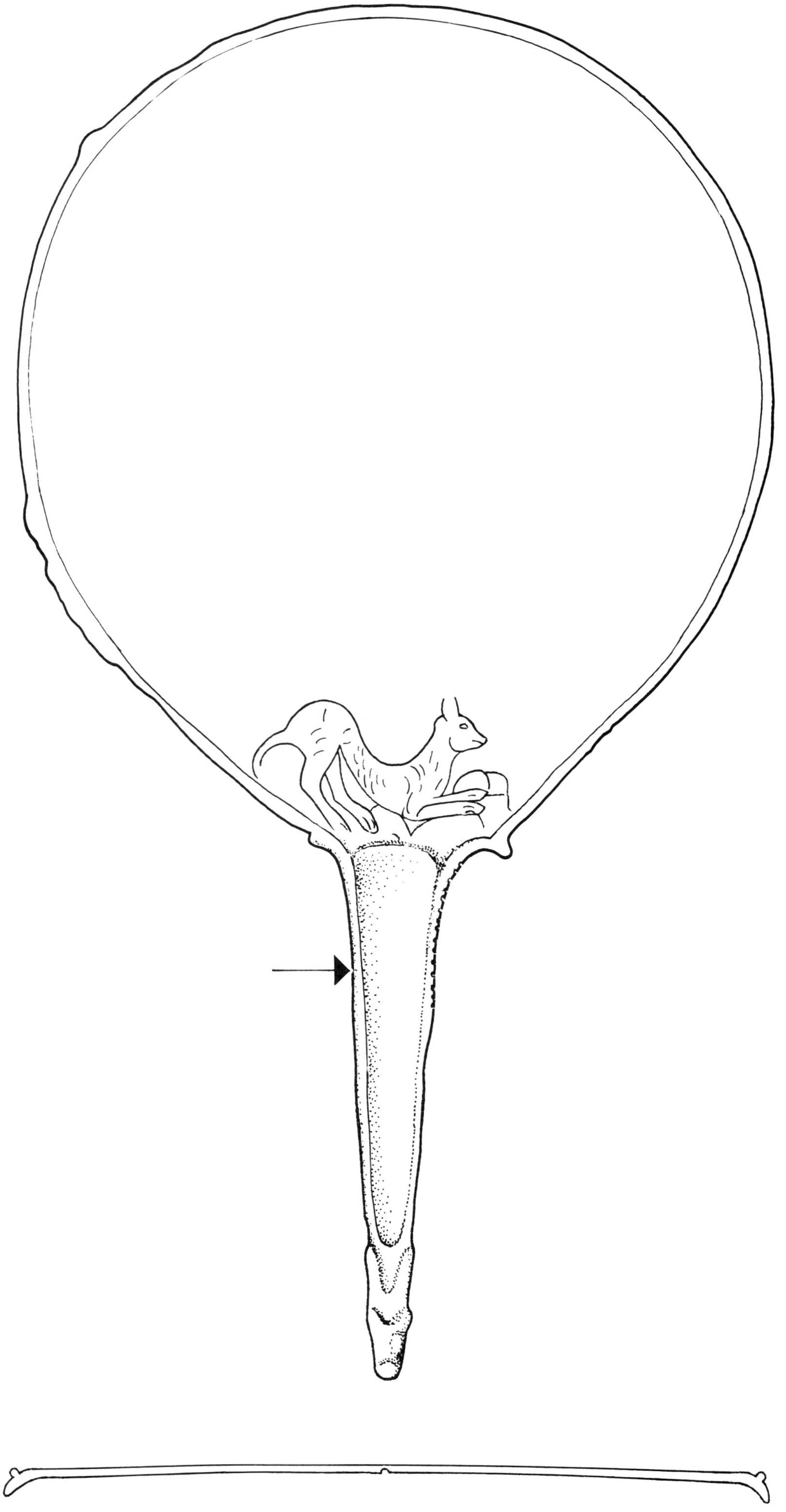

5c

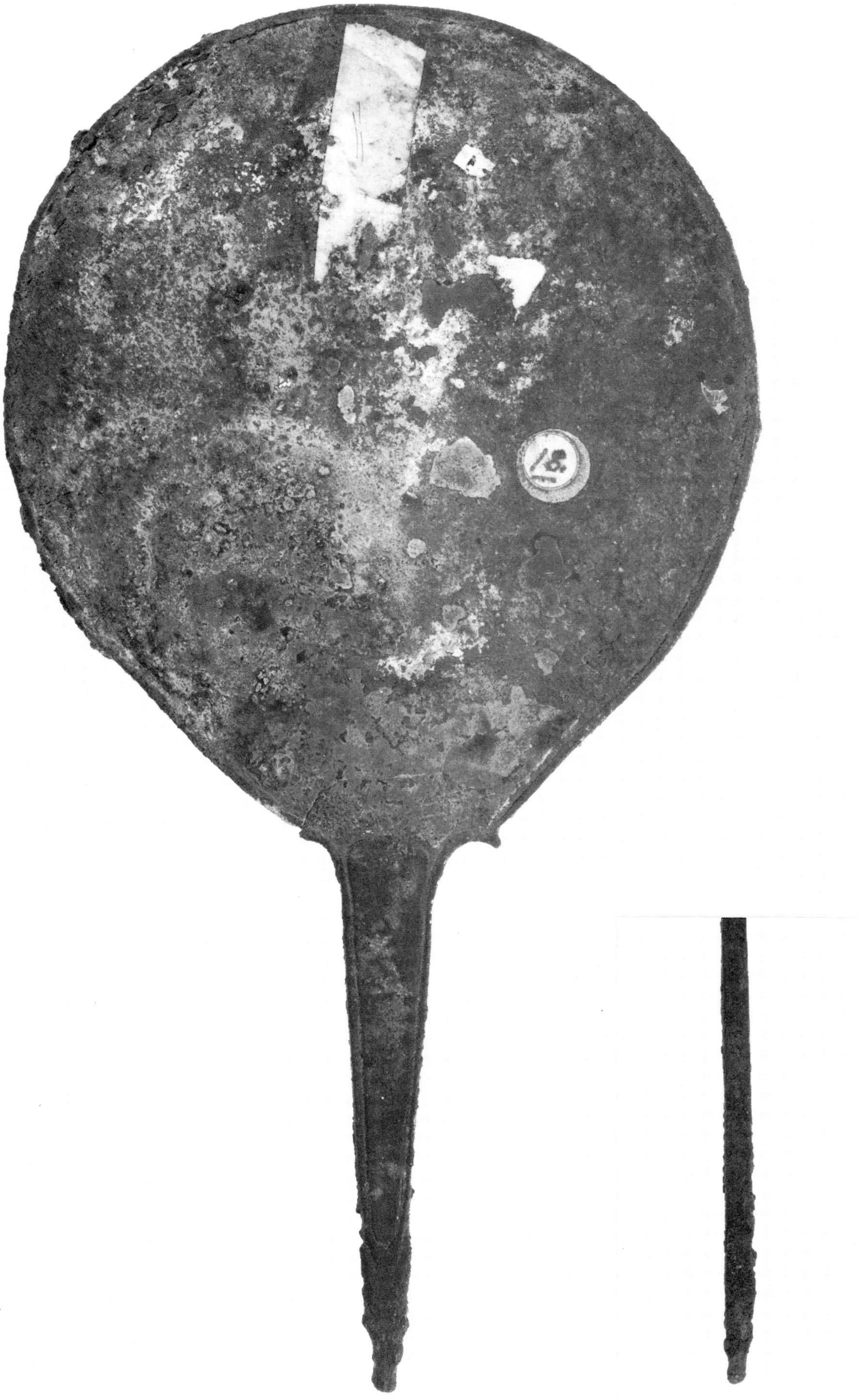

5d

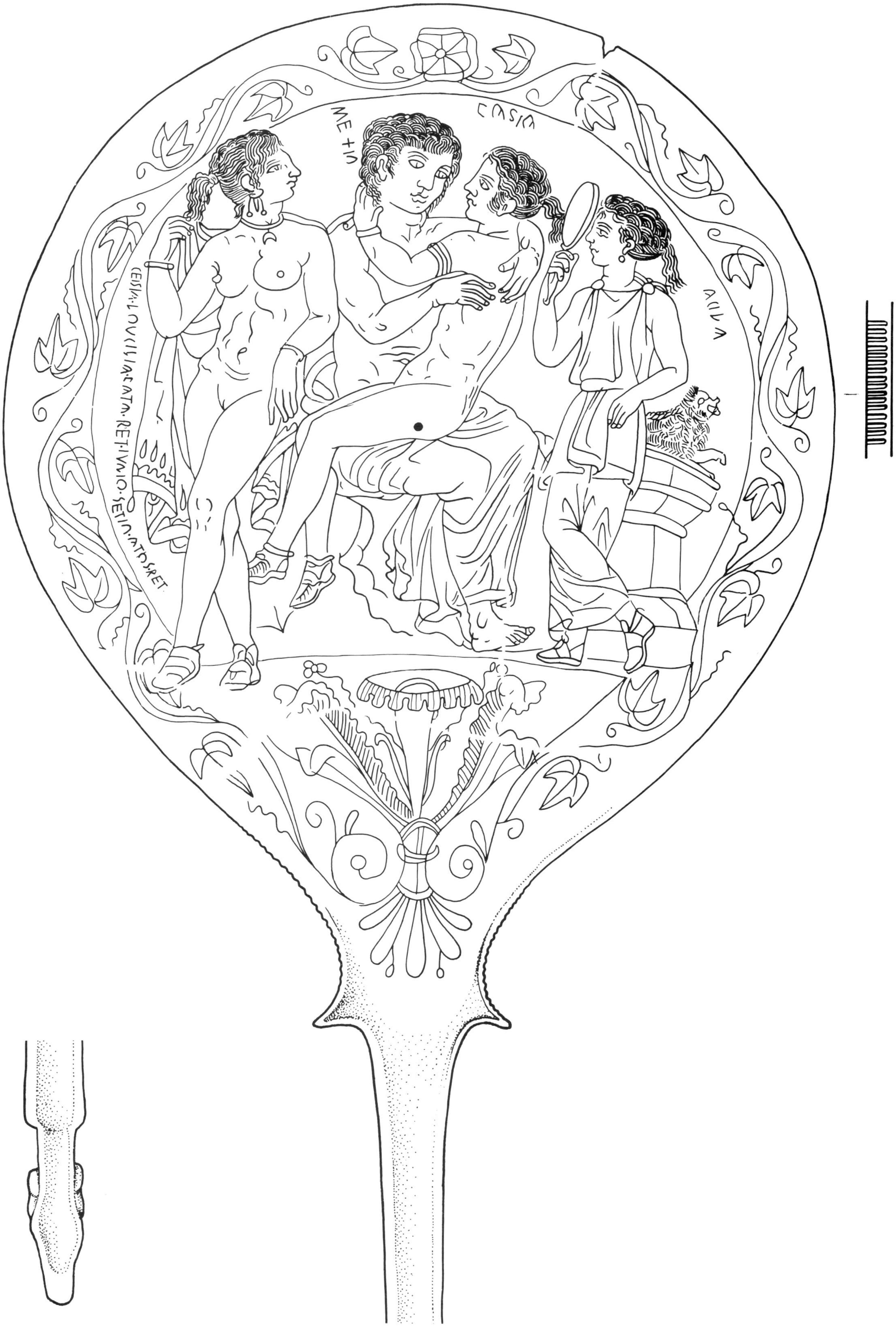

6a

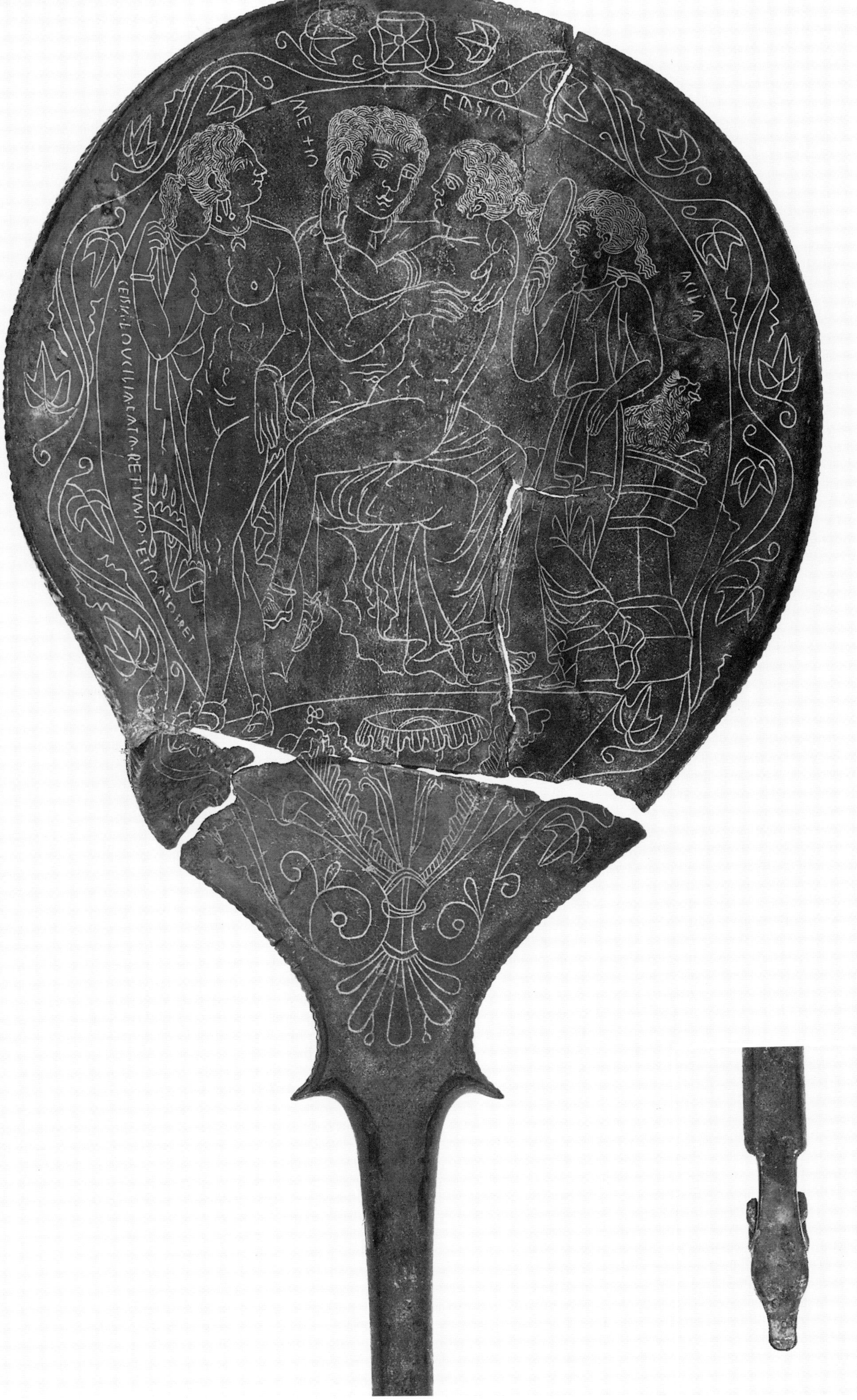

6b

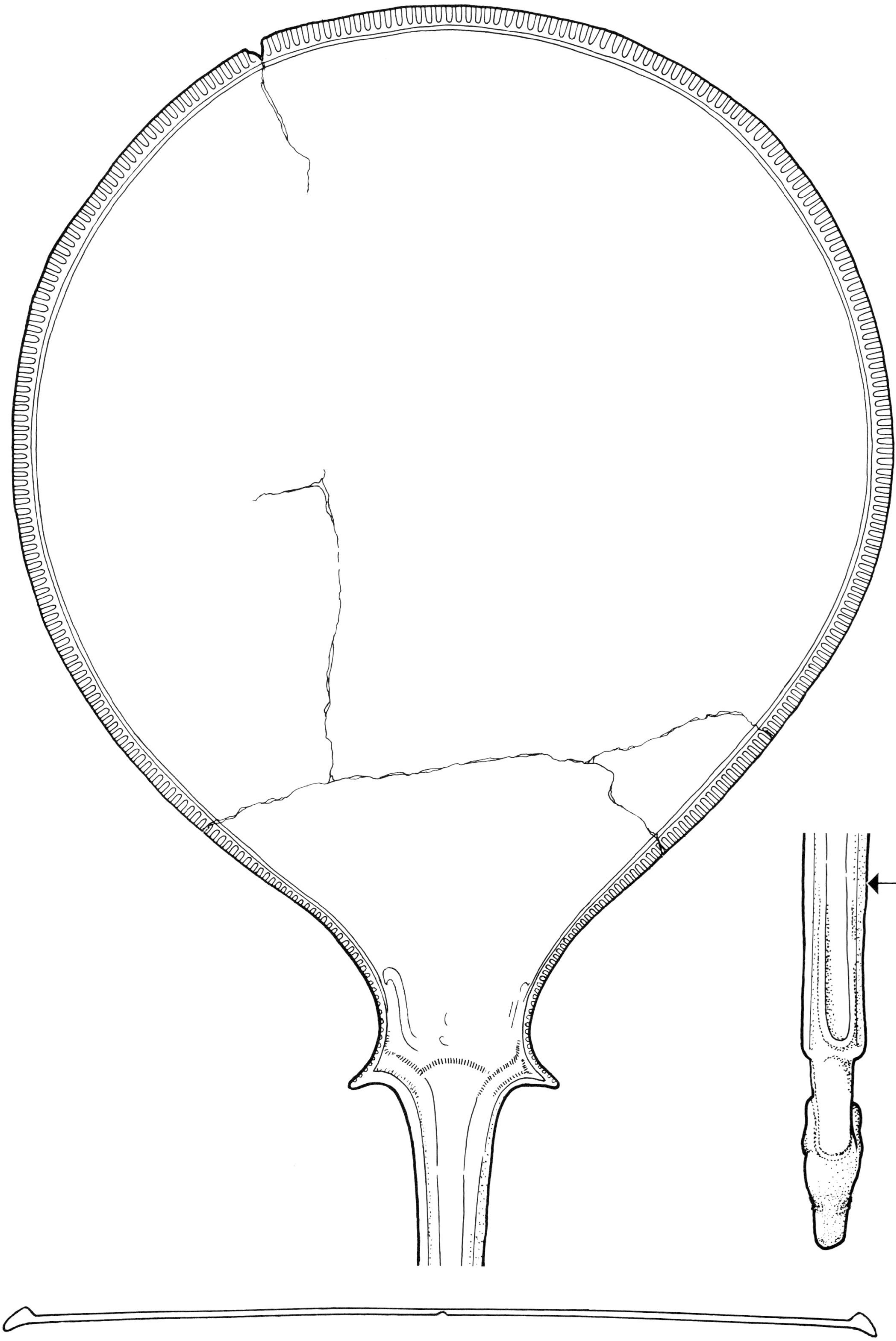

6c

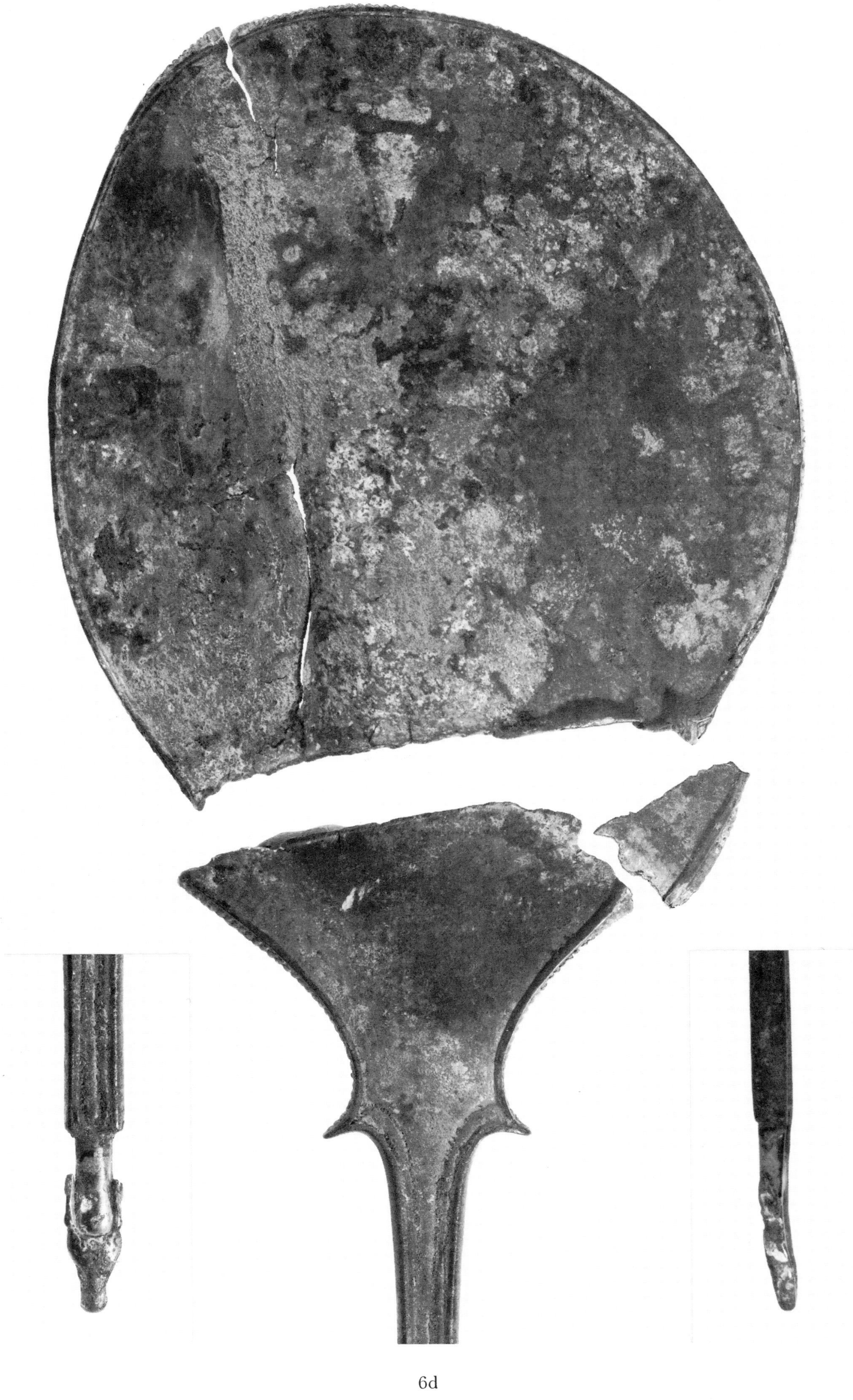

6d

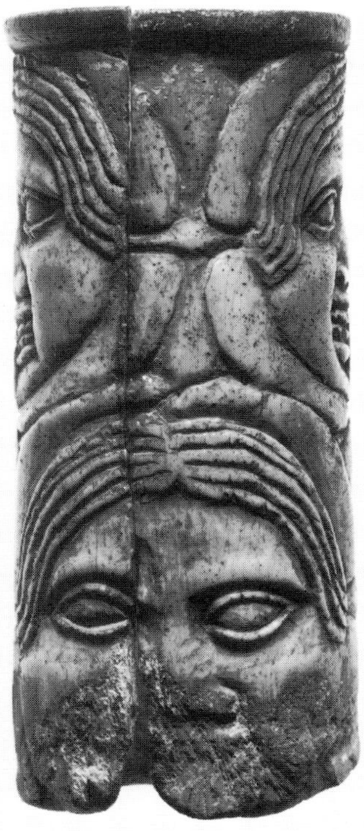

7a

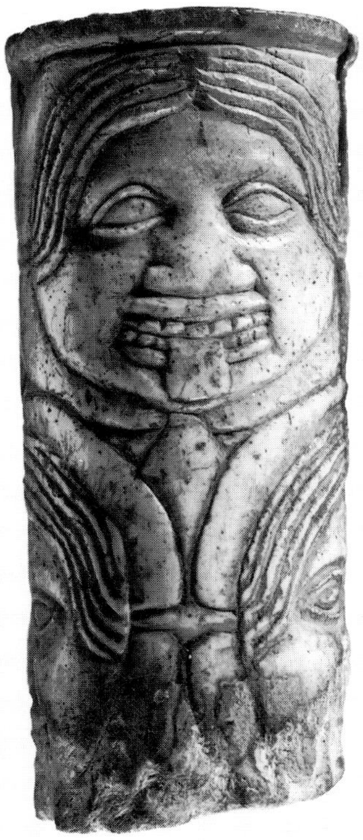

7b

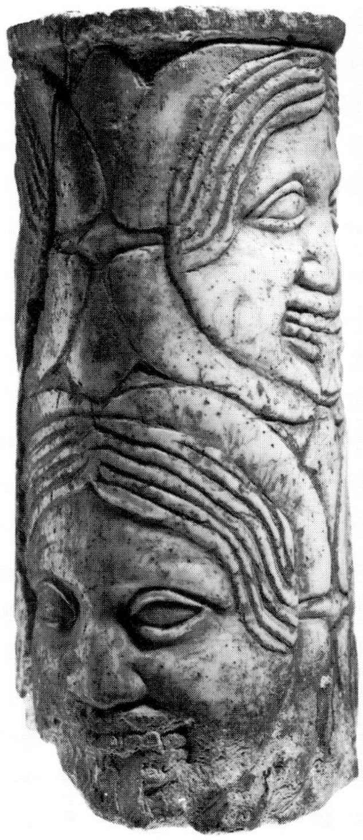

7c

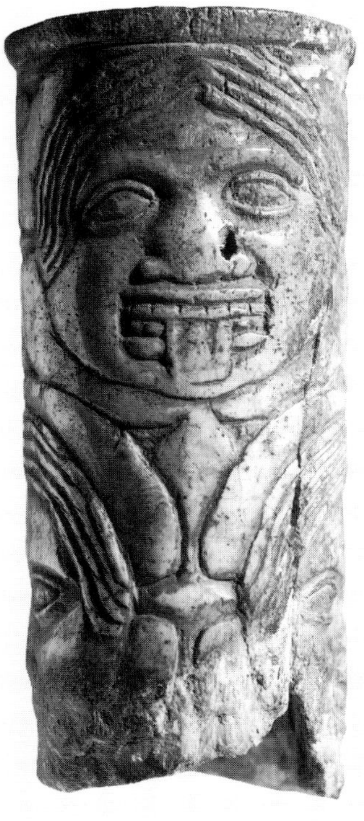

7d

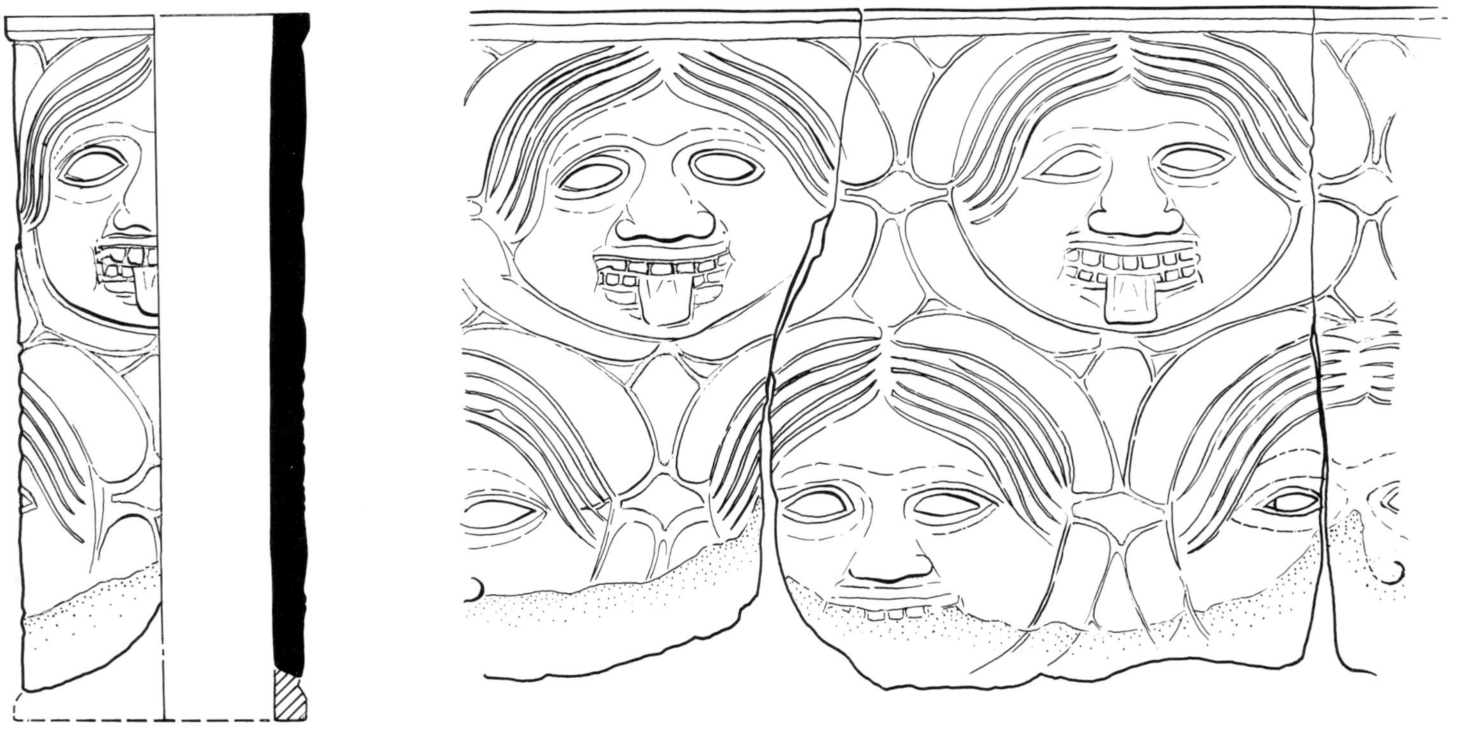

7e

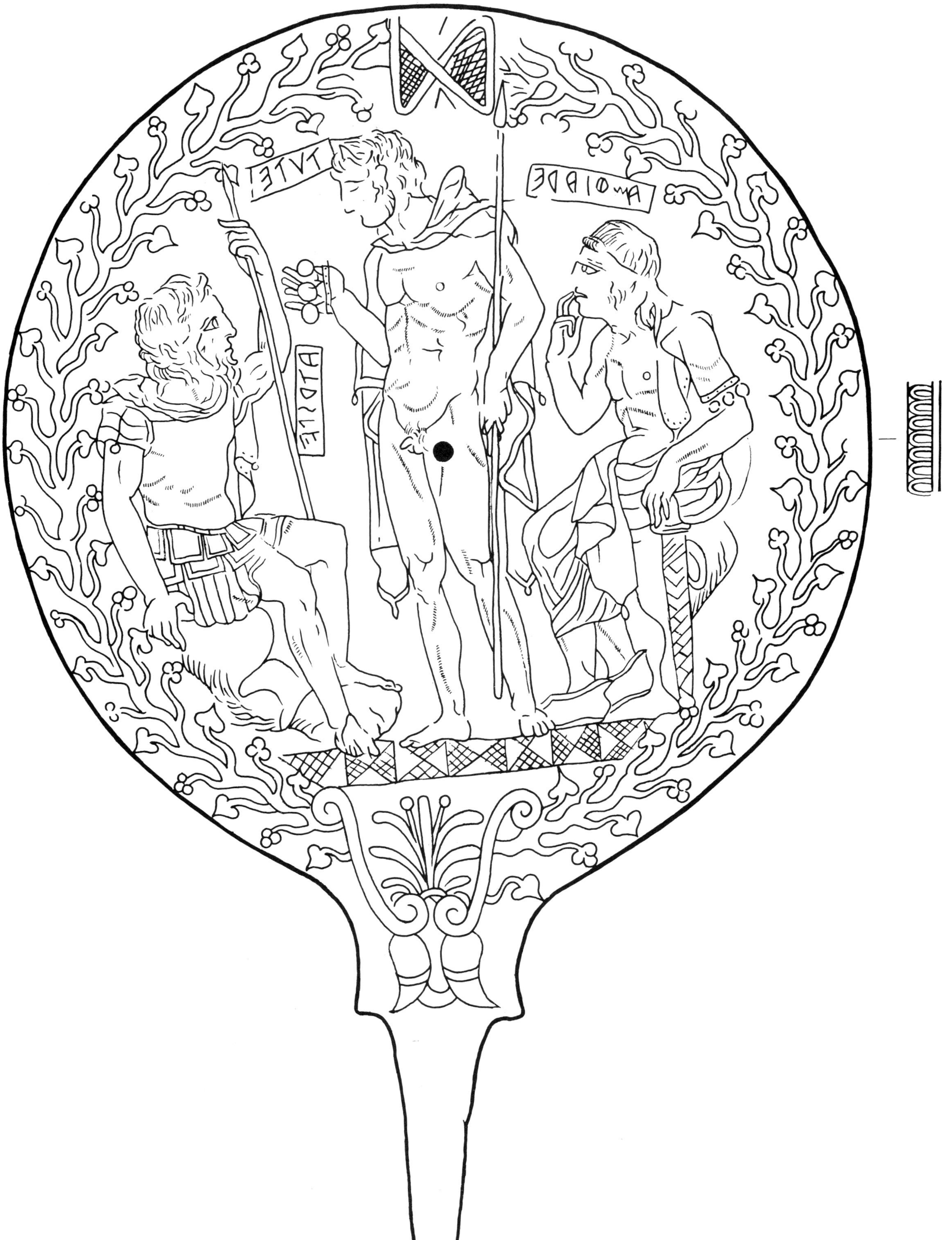

8a

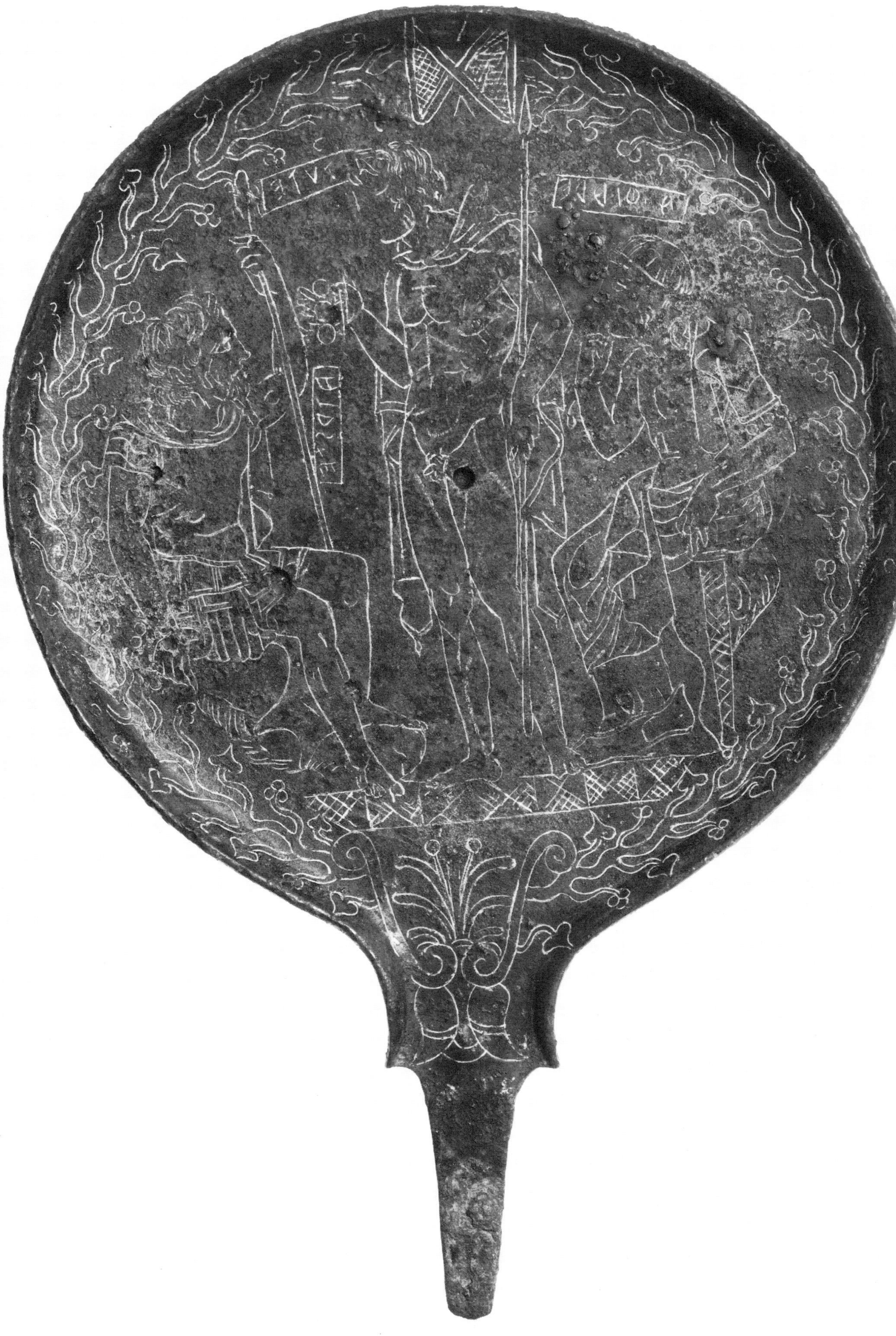

8b

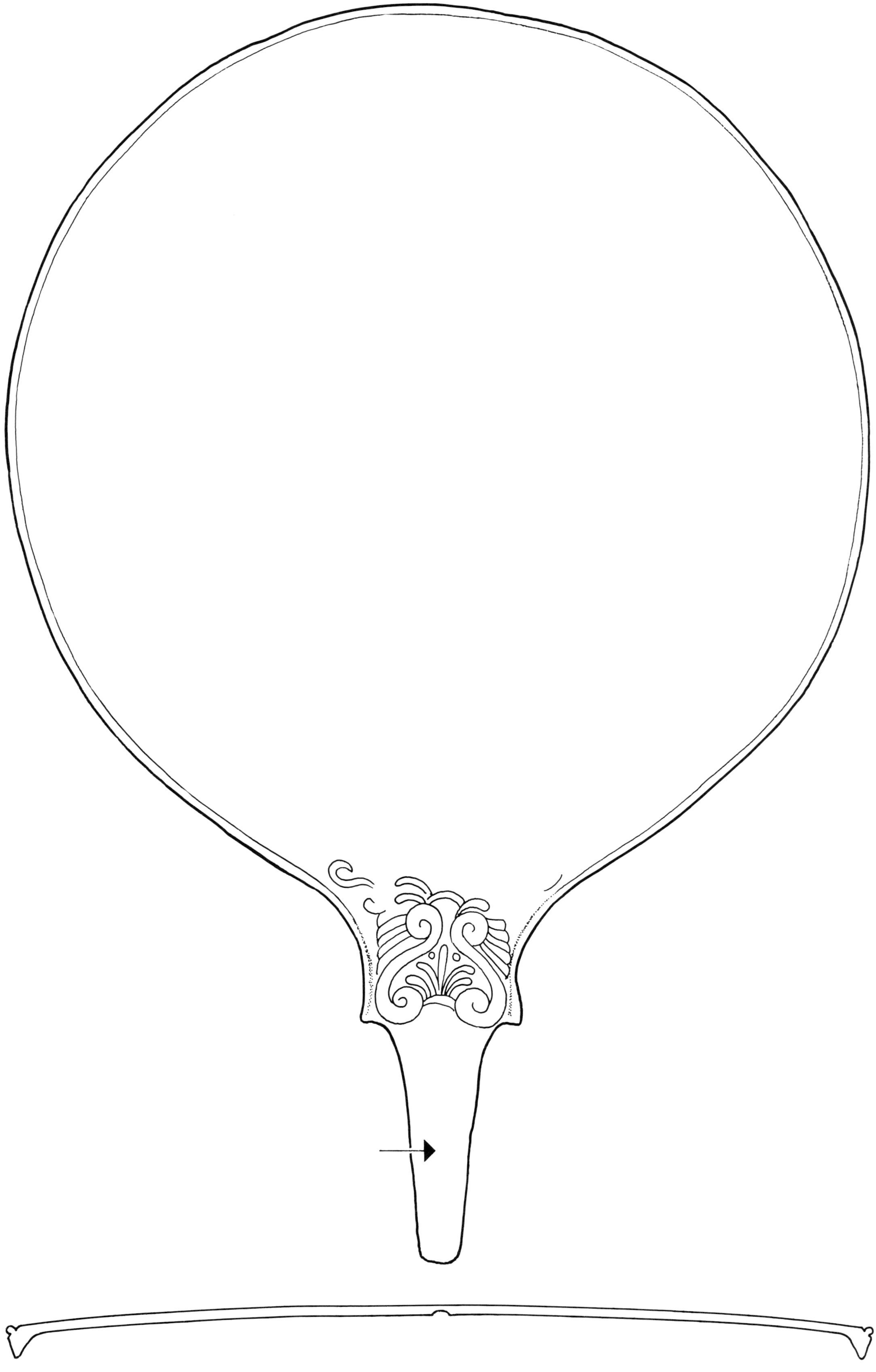

8c

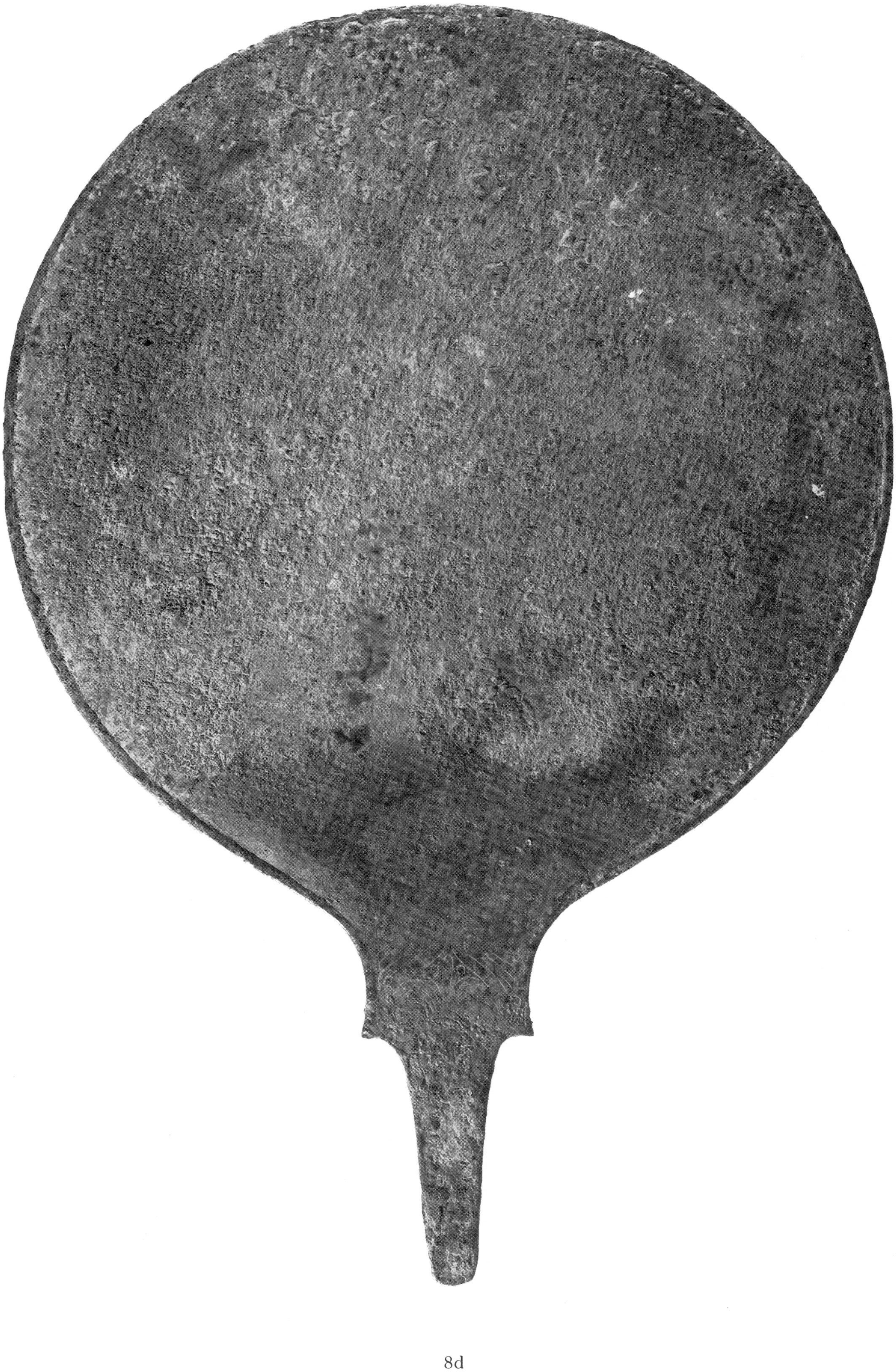

8d

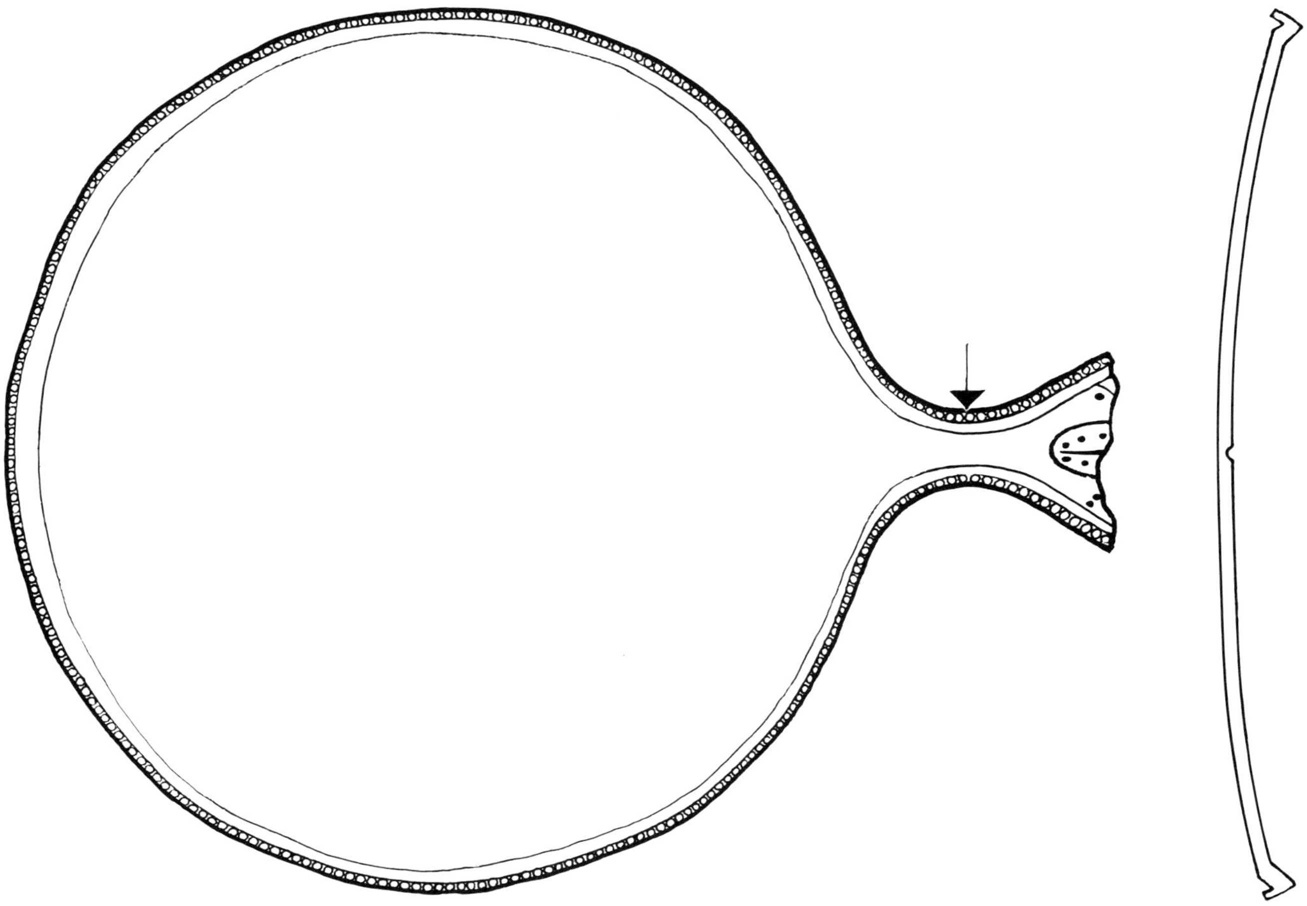

9c

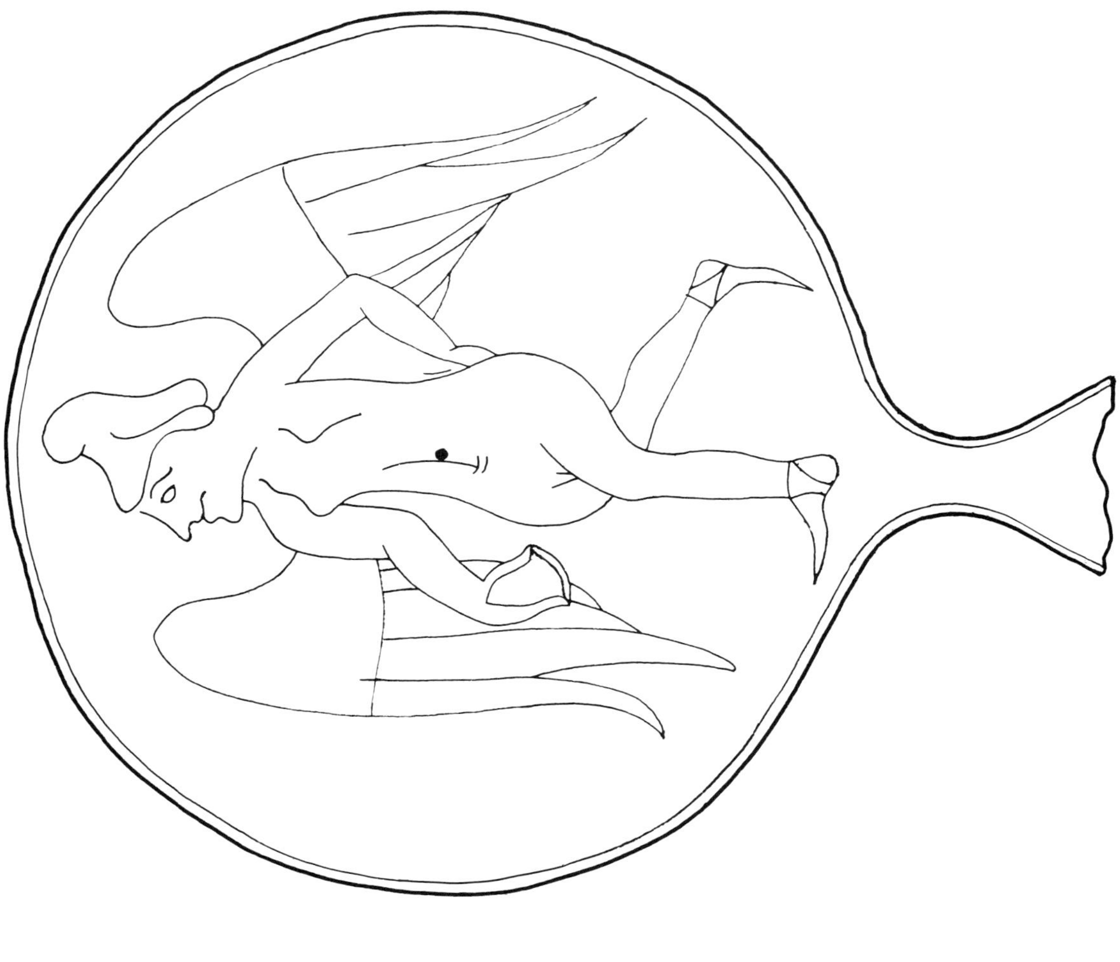

9a

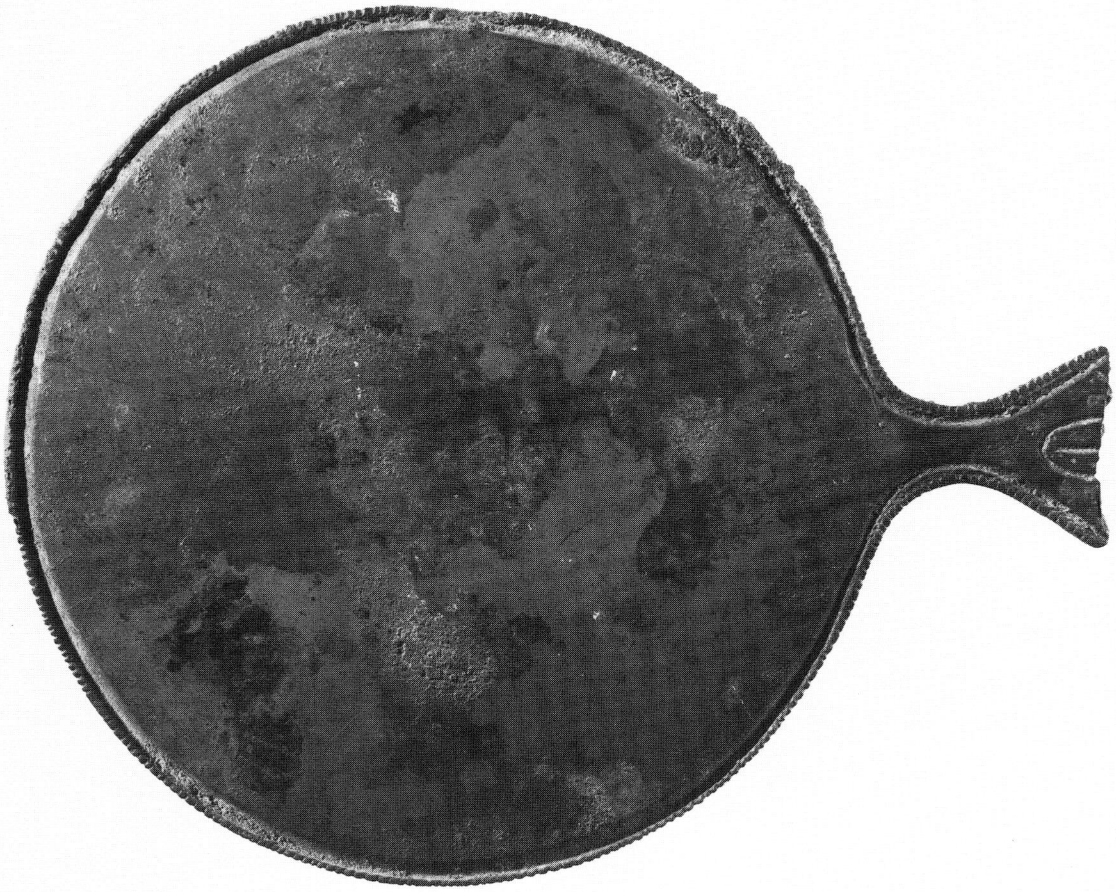

9d

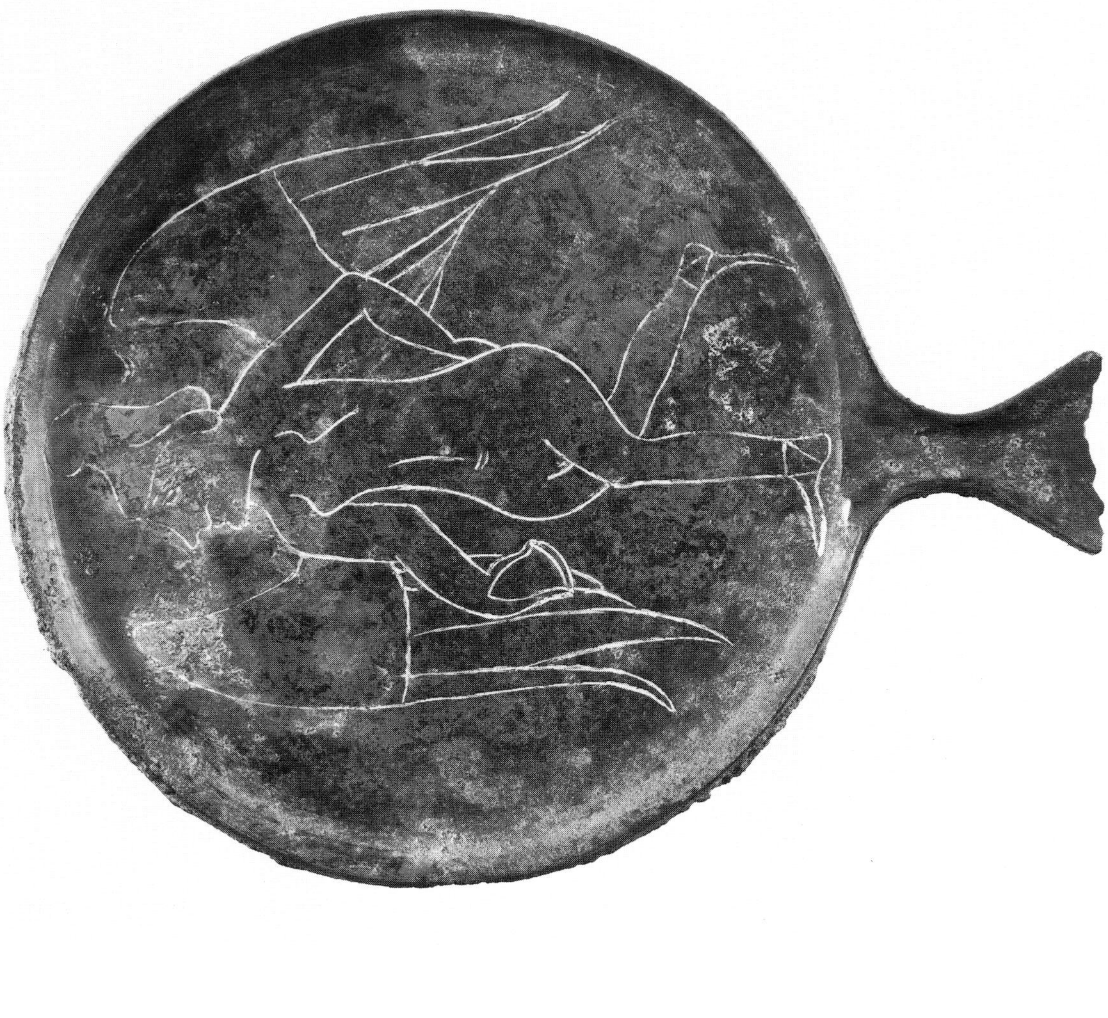

9b

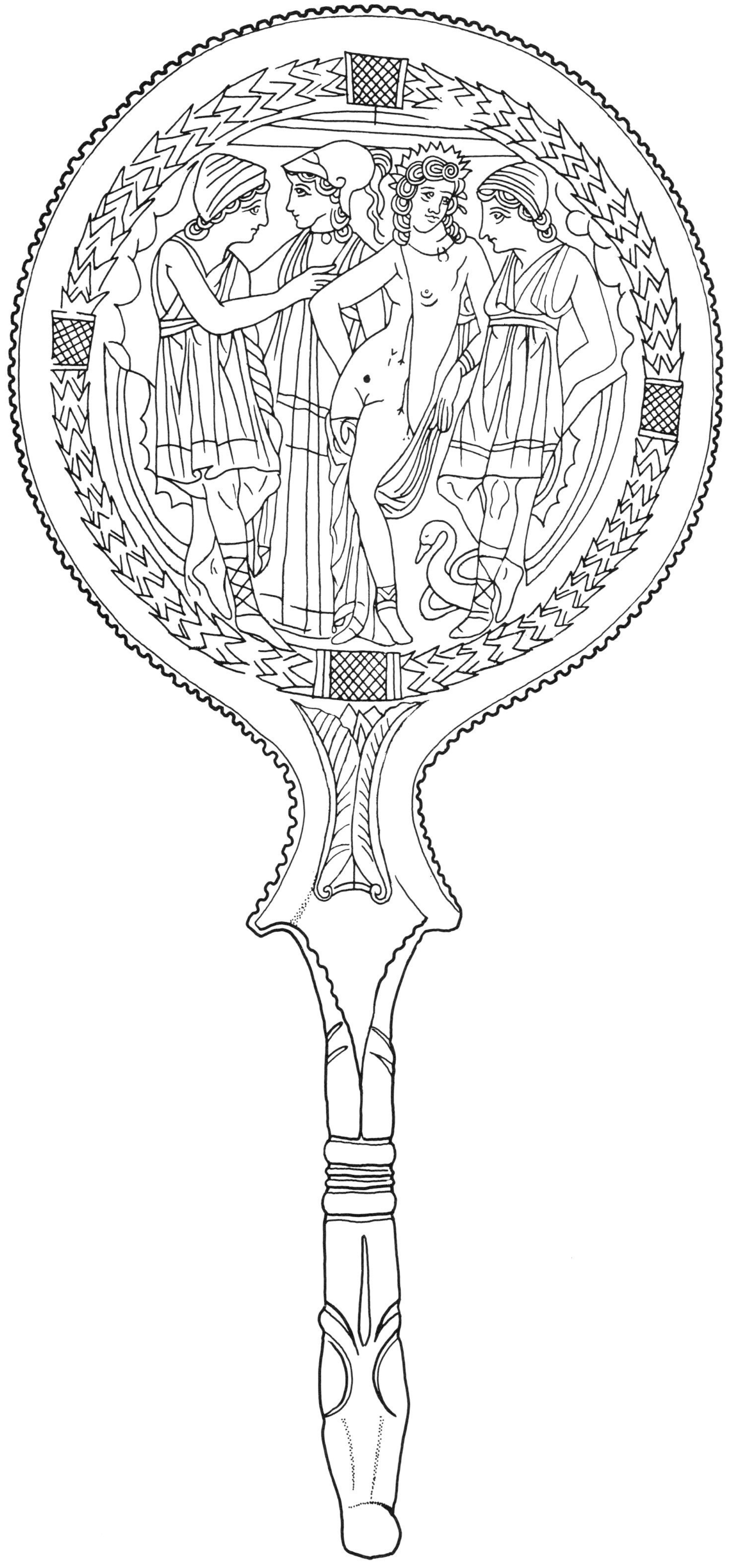

10a

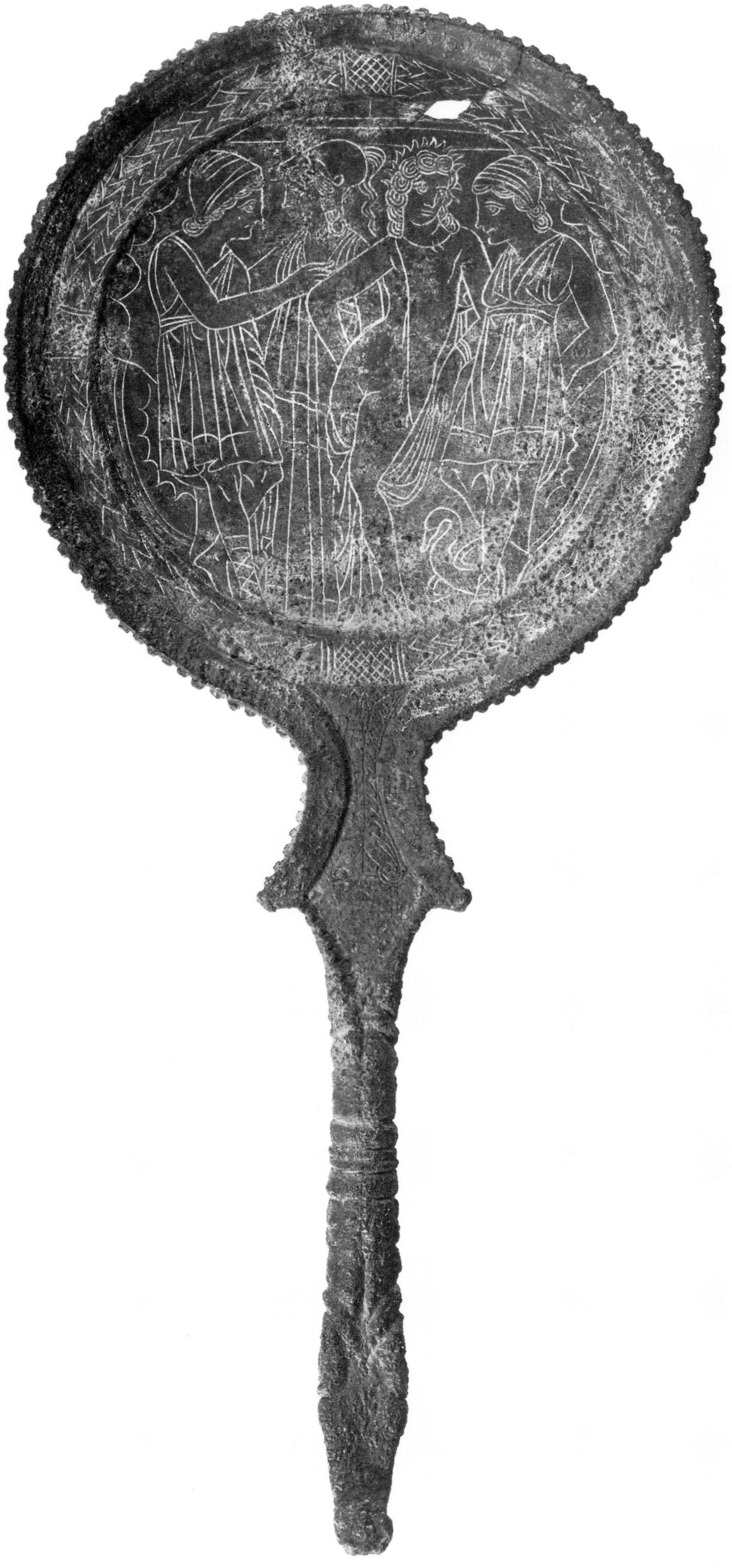

10b

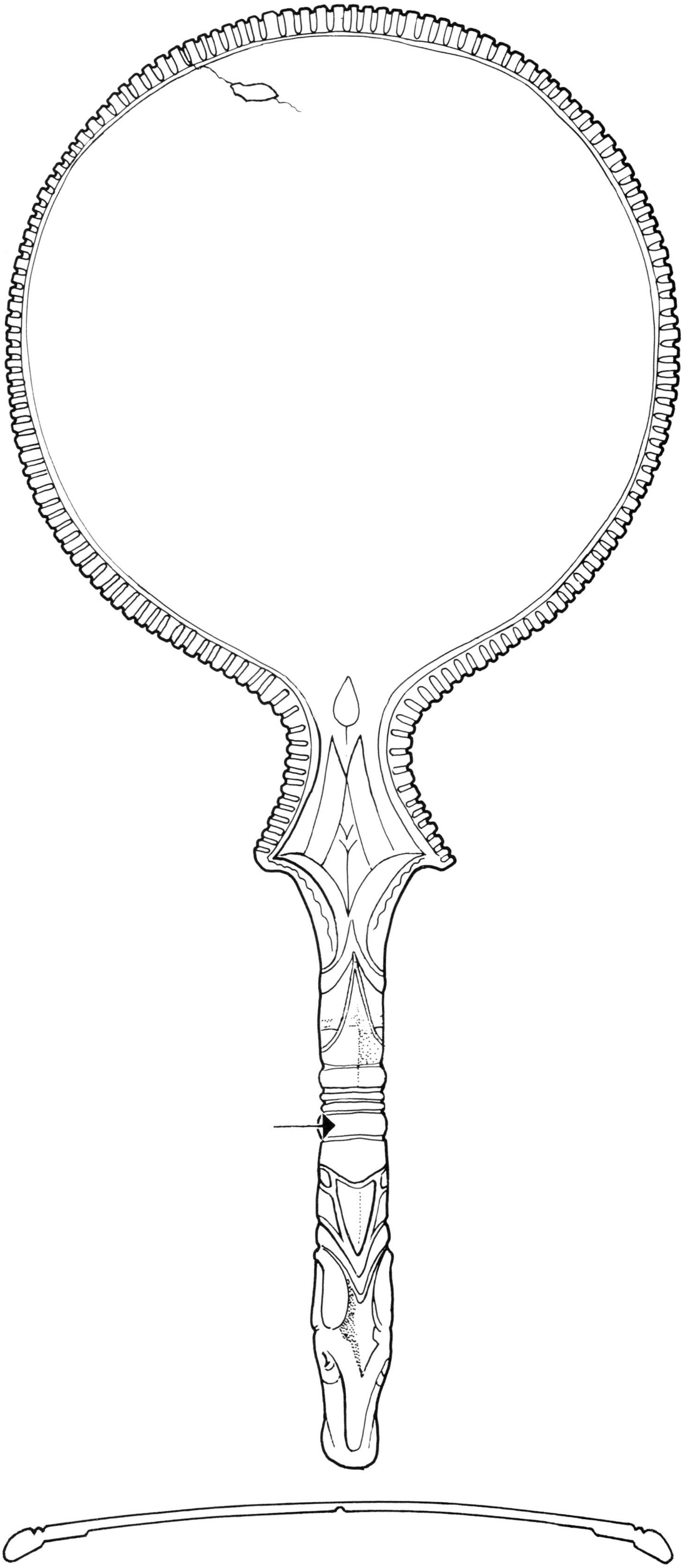

10c

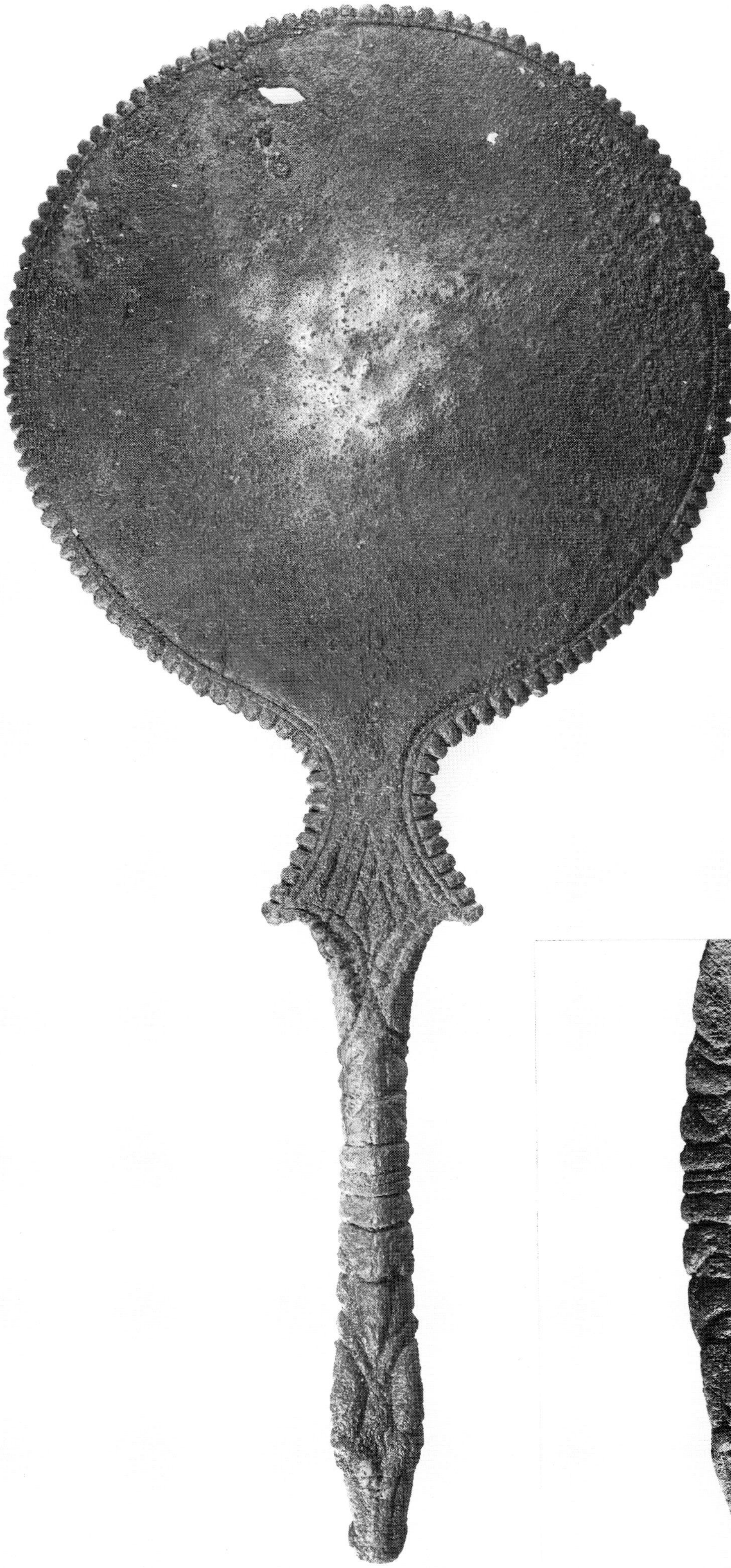

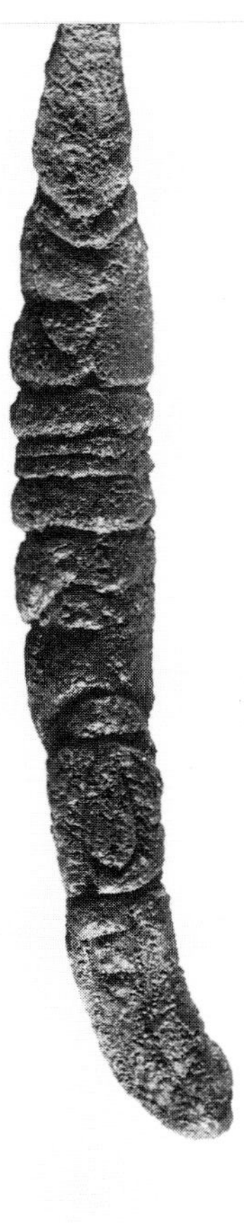

10d

CSE,
GREAT
BRITAIN
2, 11

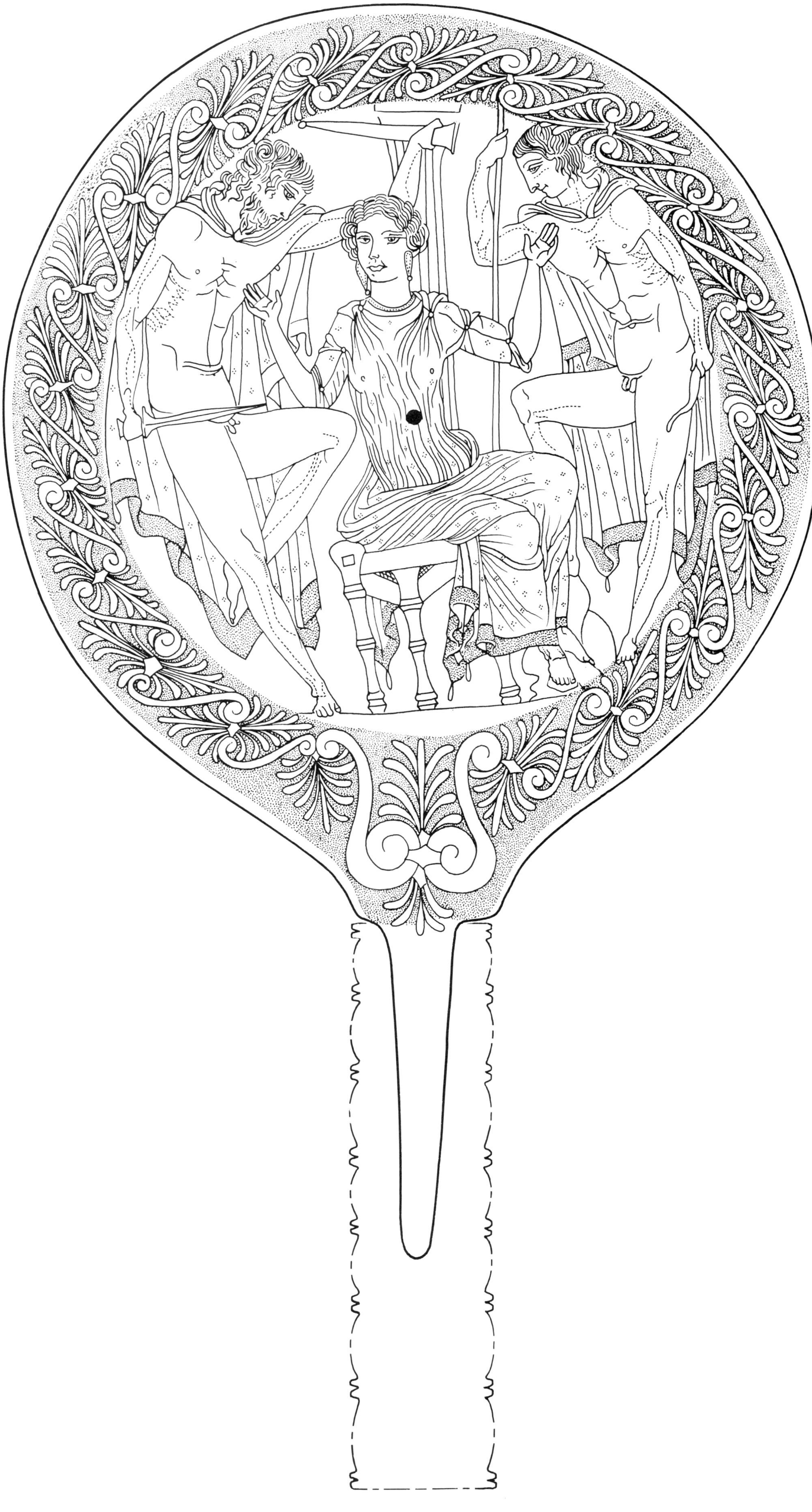

11a

CSE,
GREAT
BRITAIN
2, 11

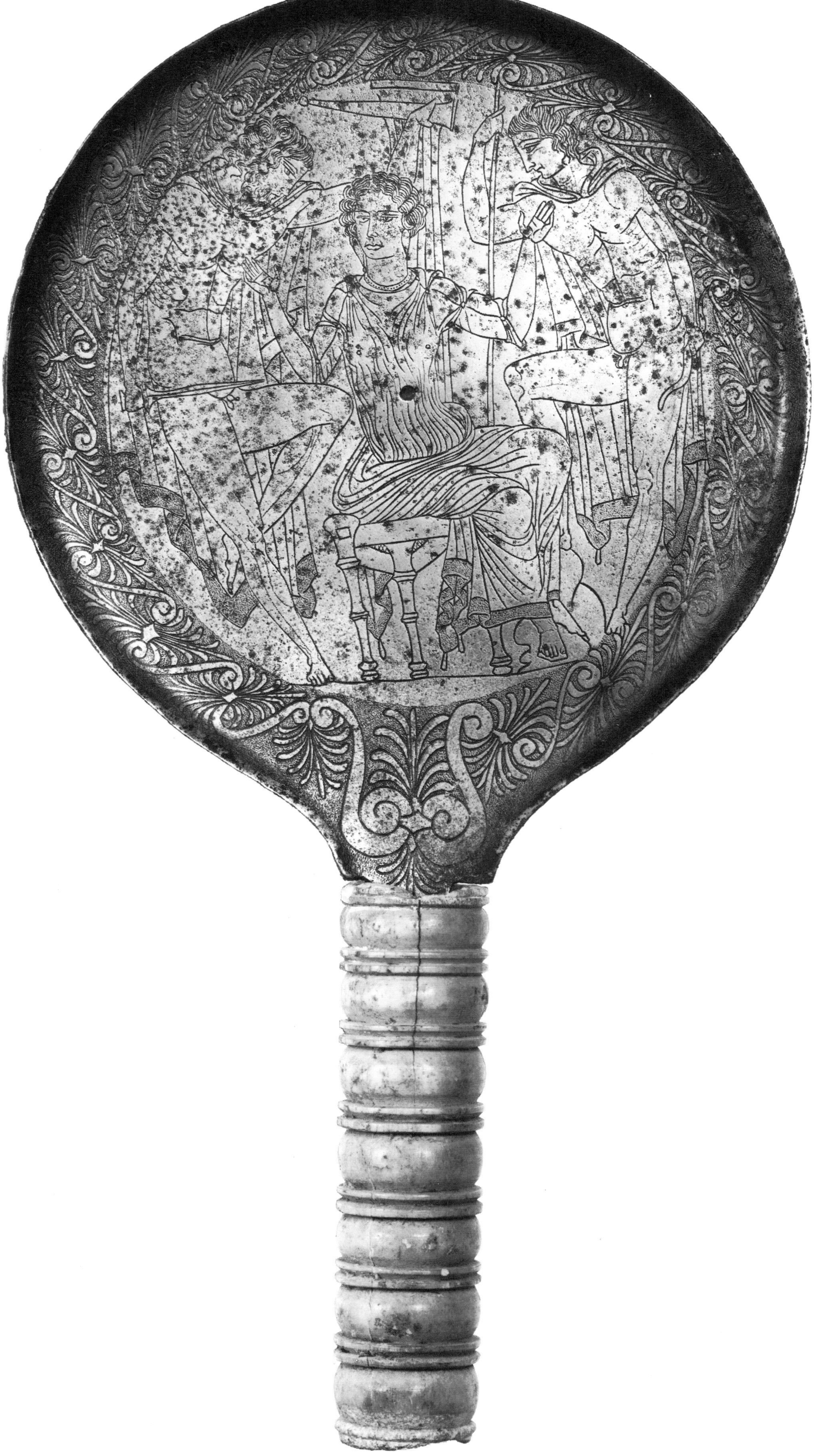

11b

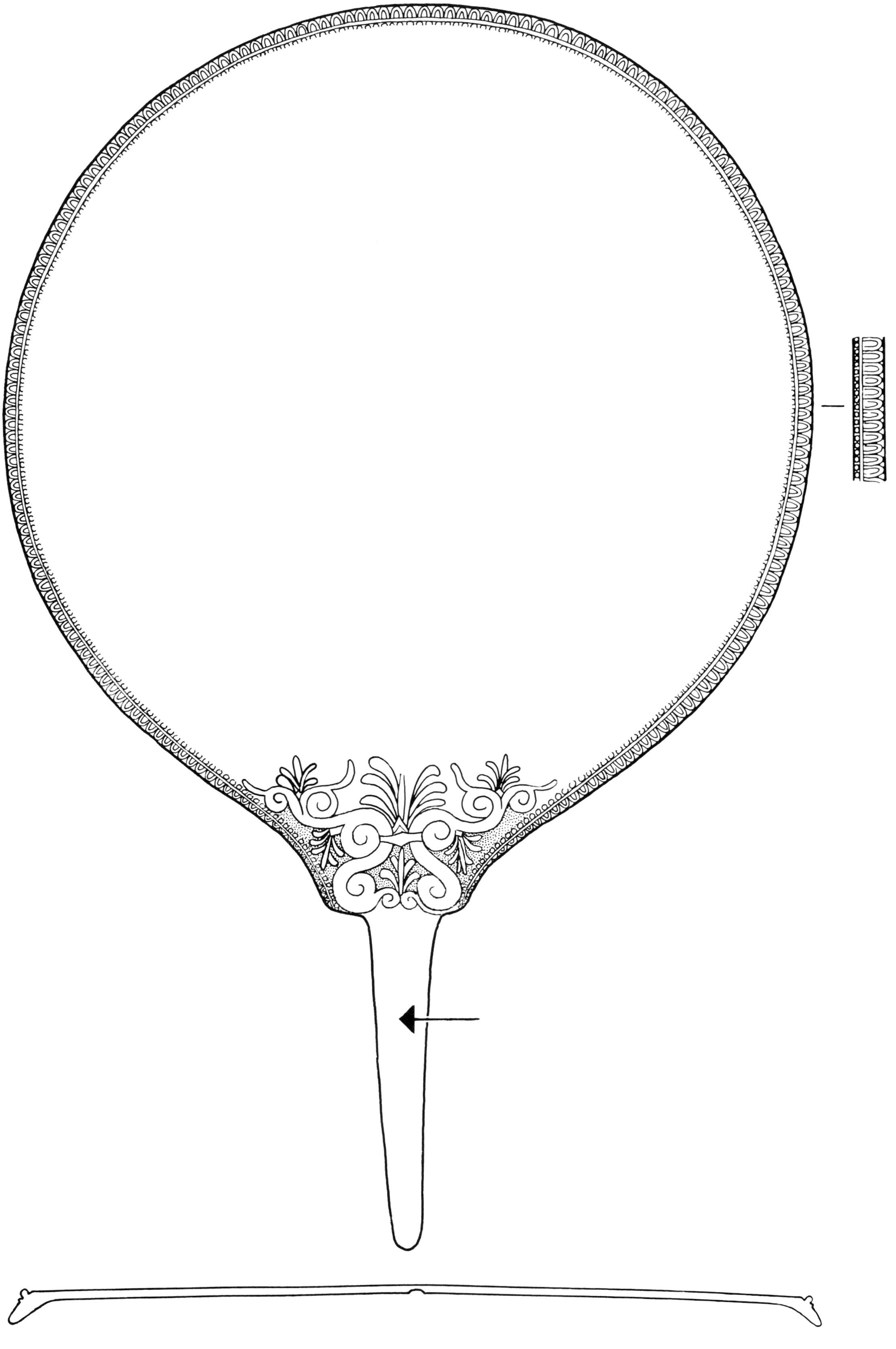

11c

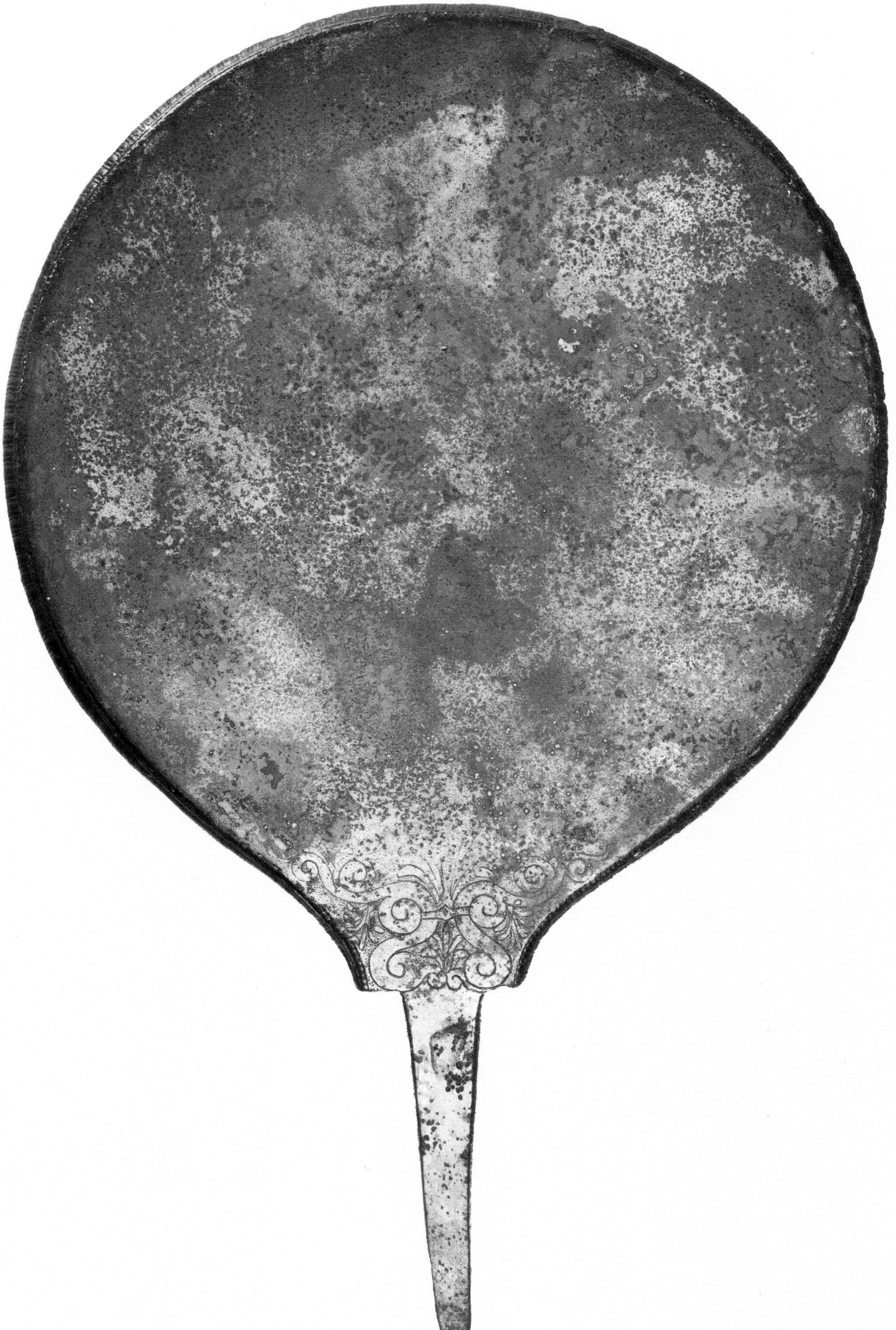

11d

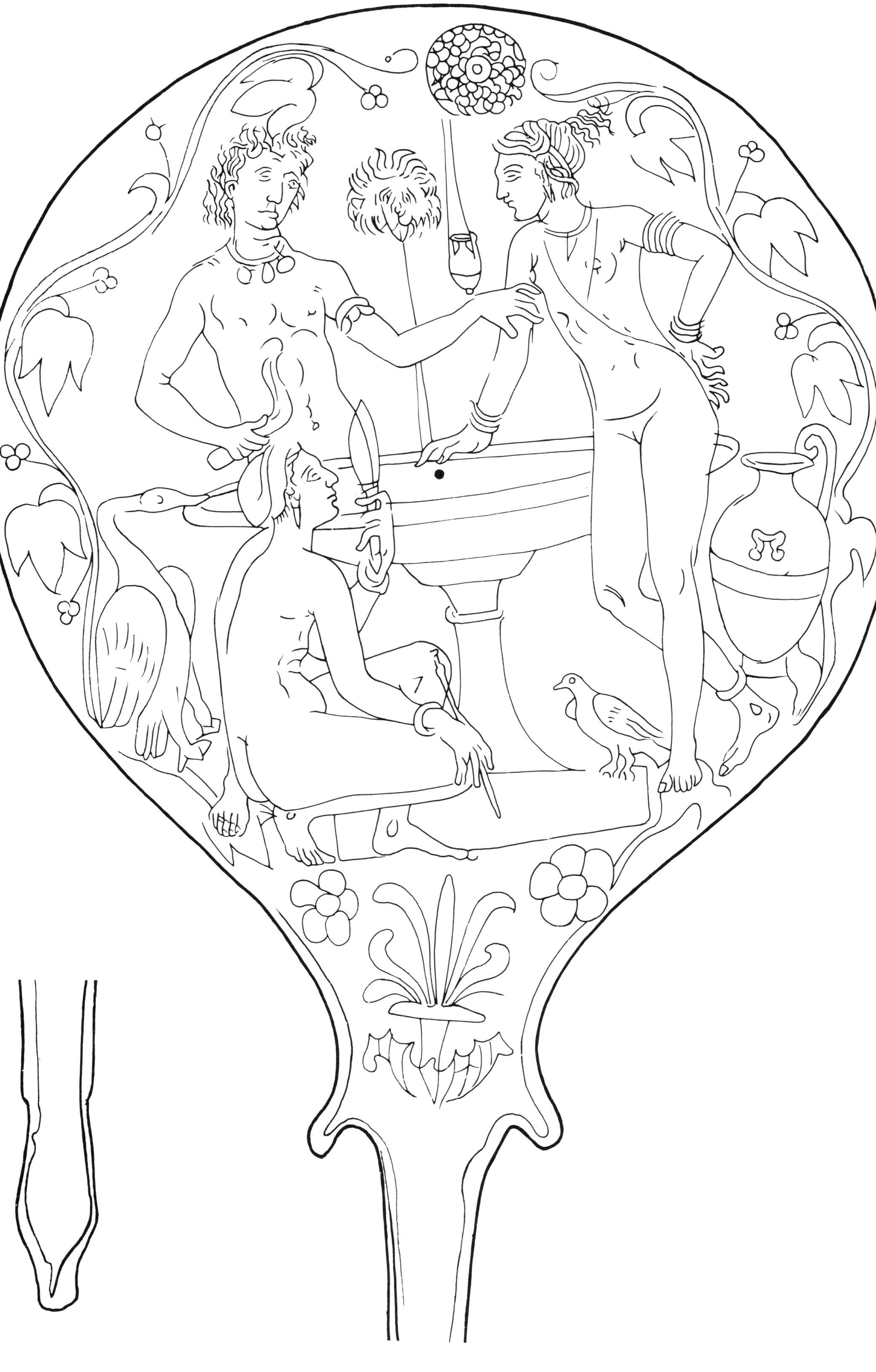

12a

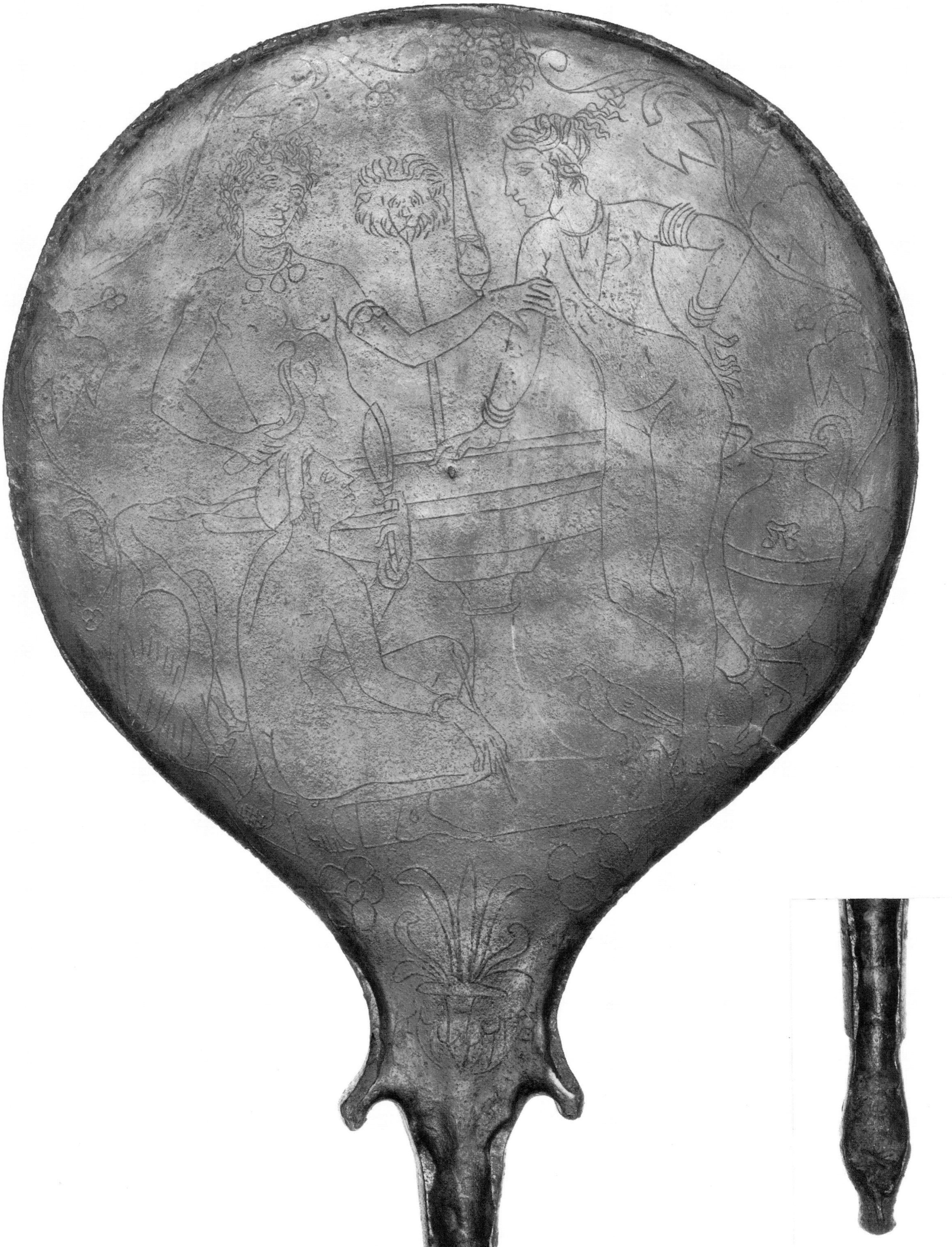

12b

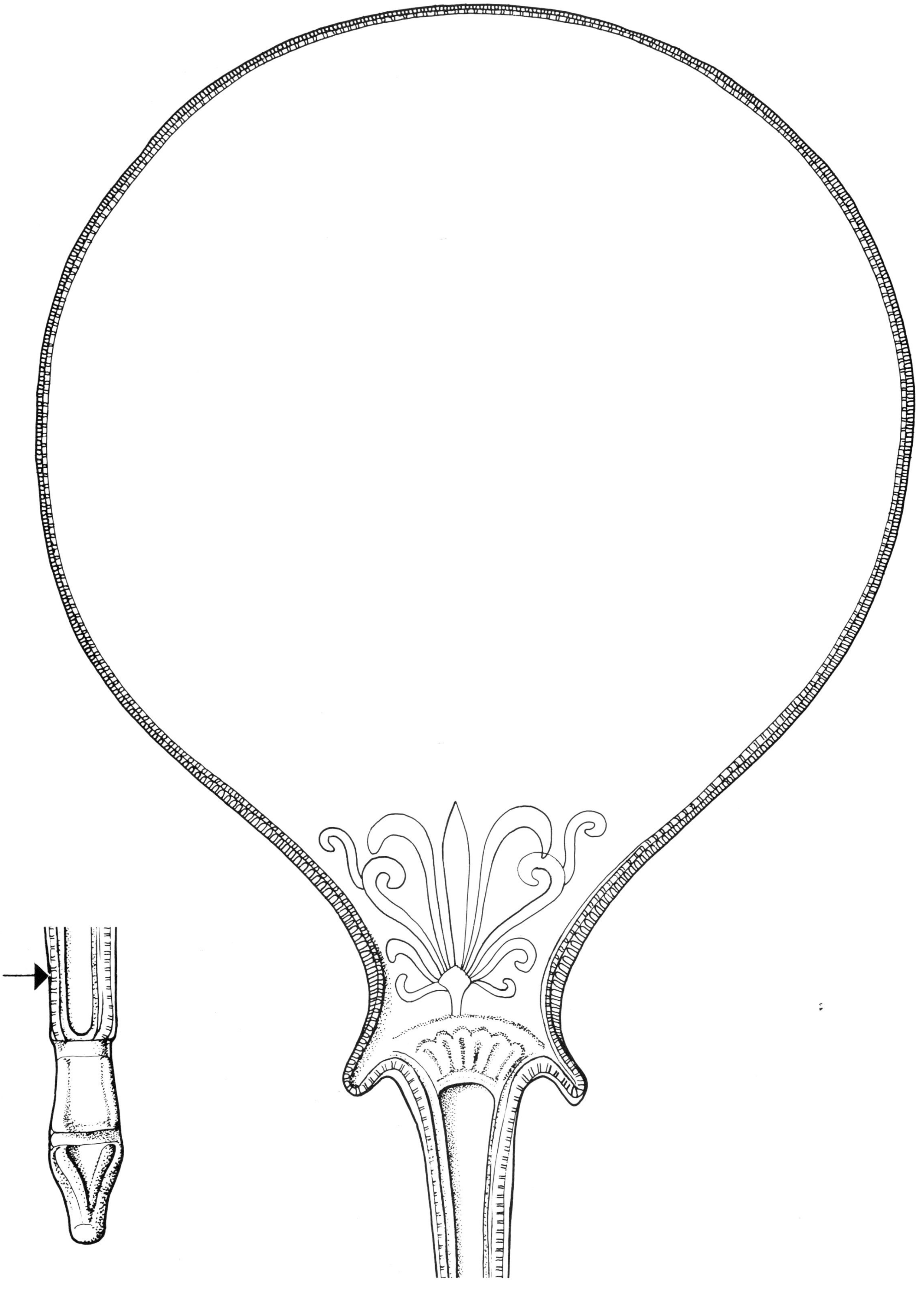

12c

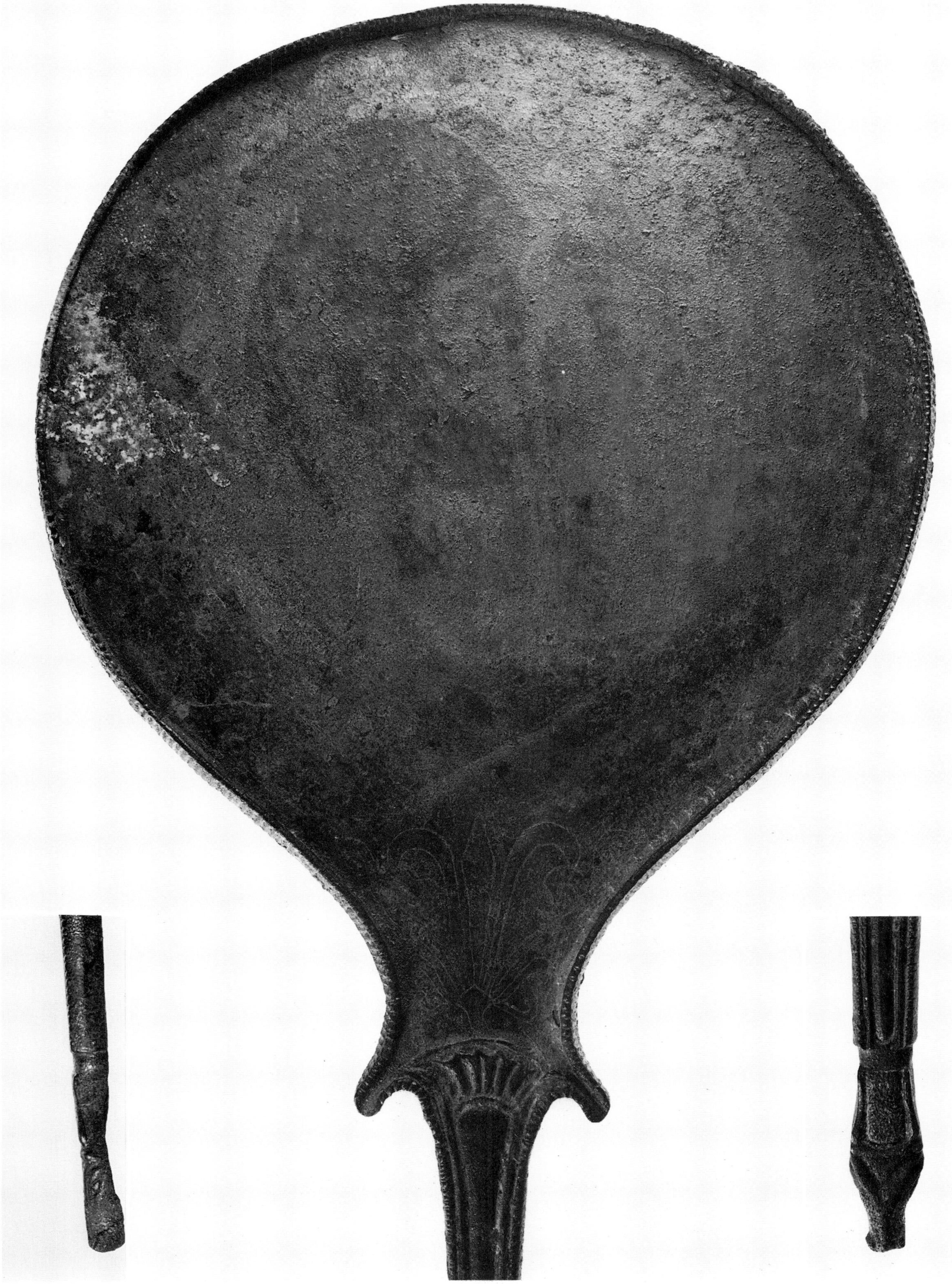

12d

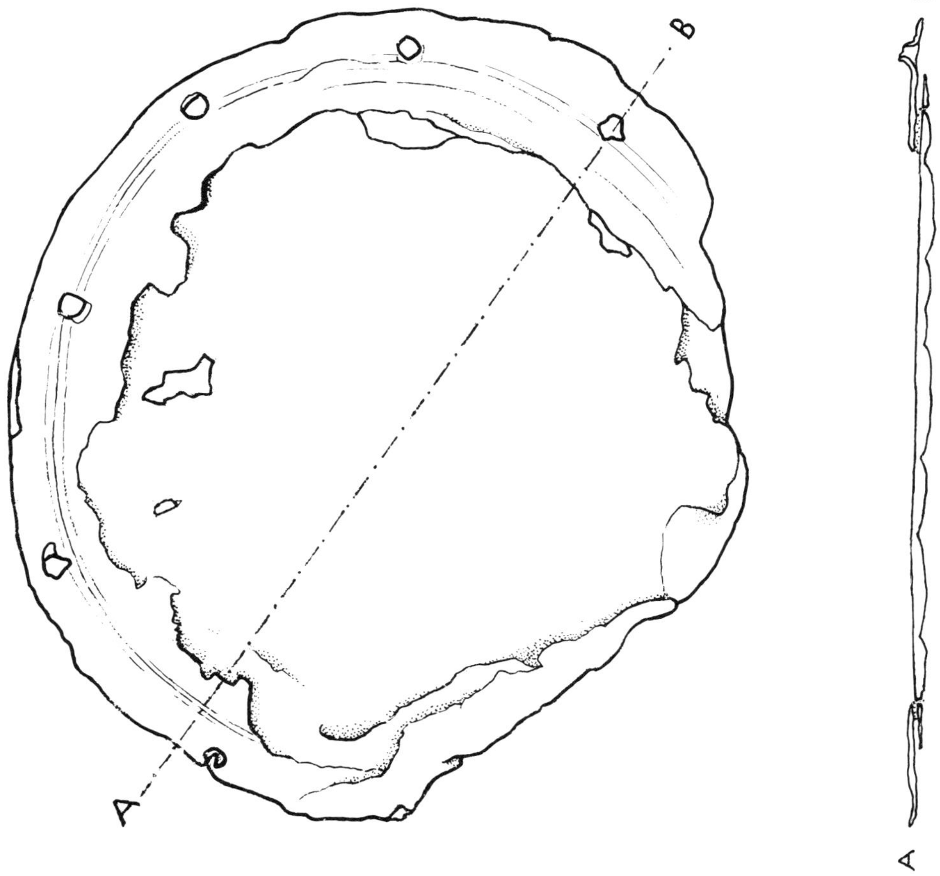

13c

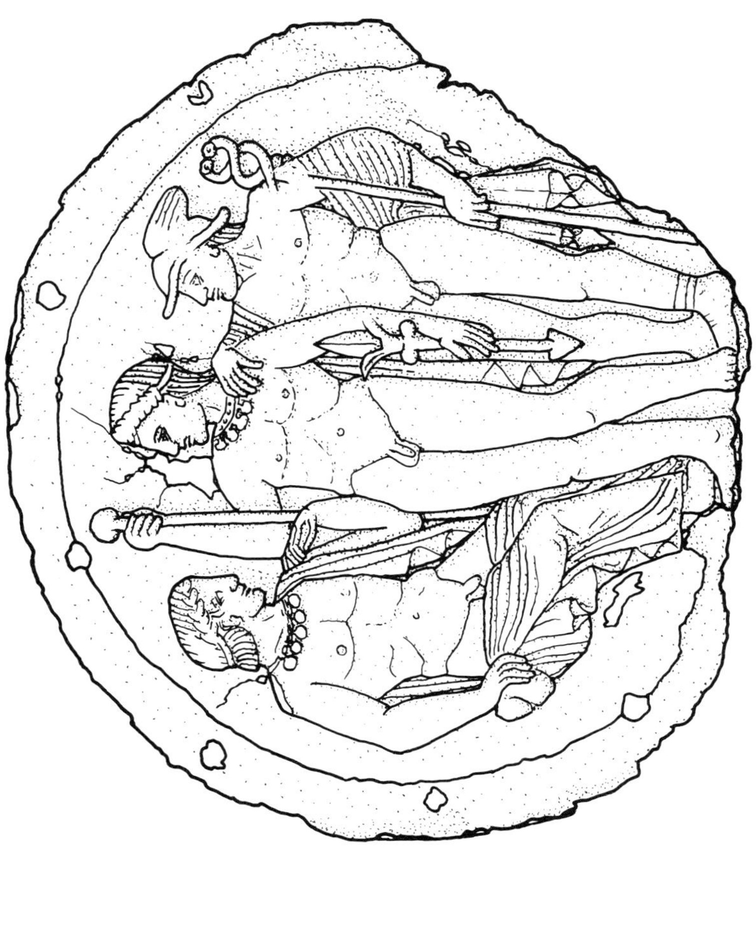

13a

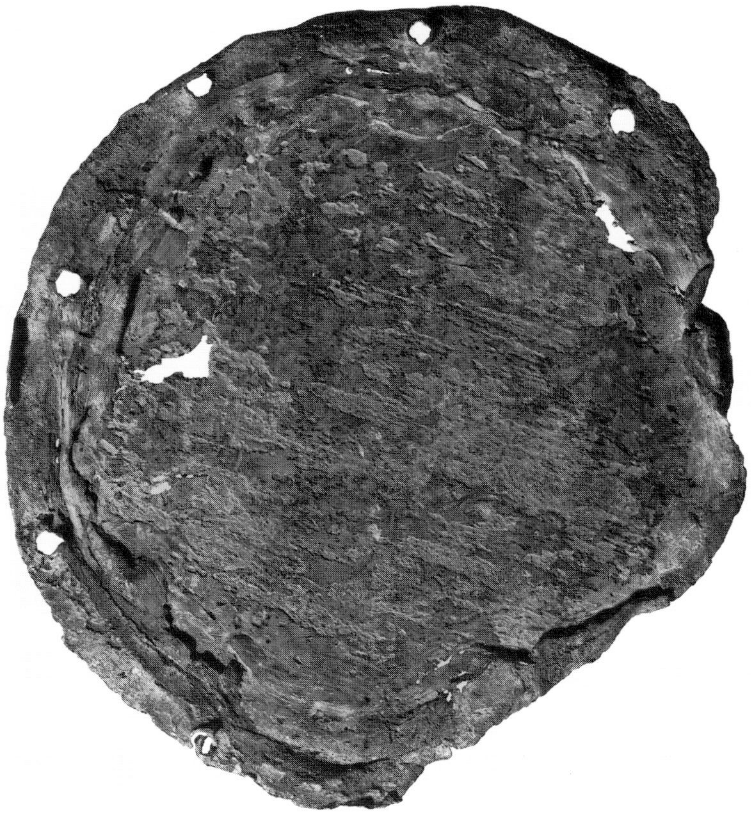

13d

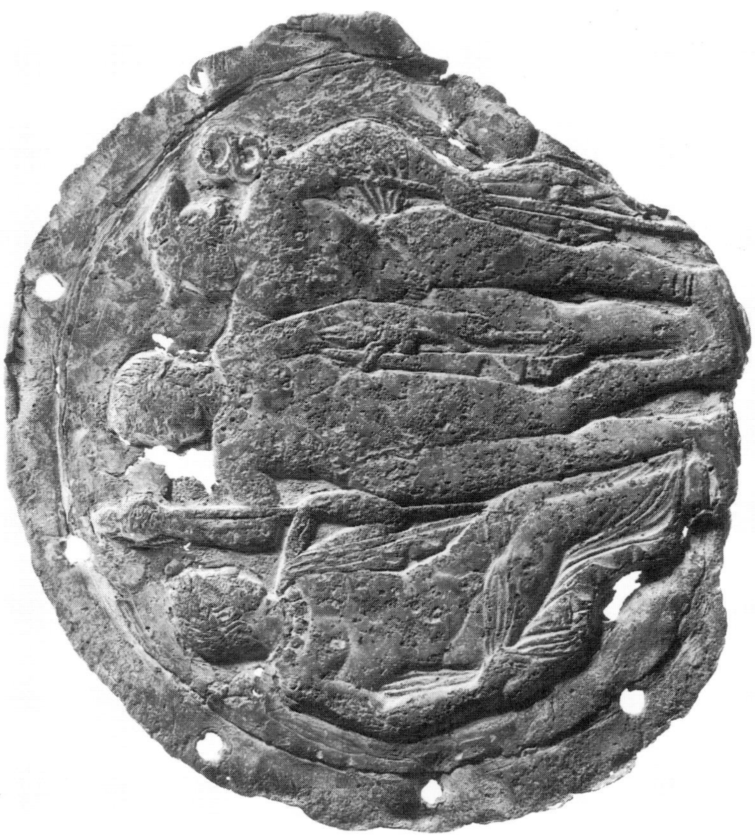

13b

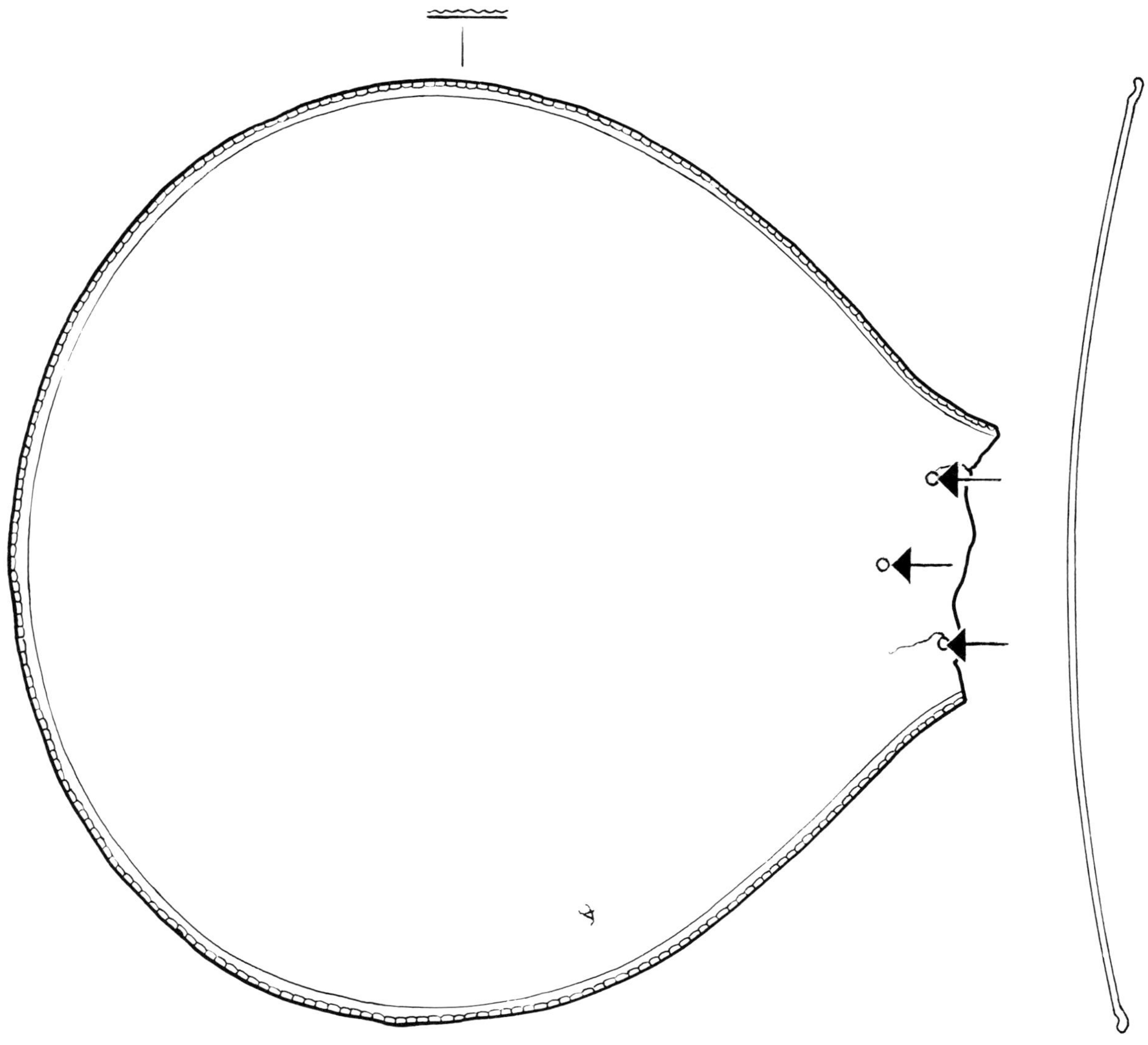

14c

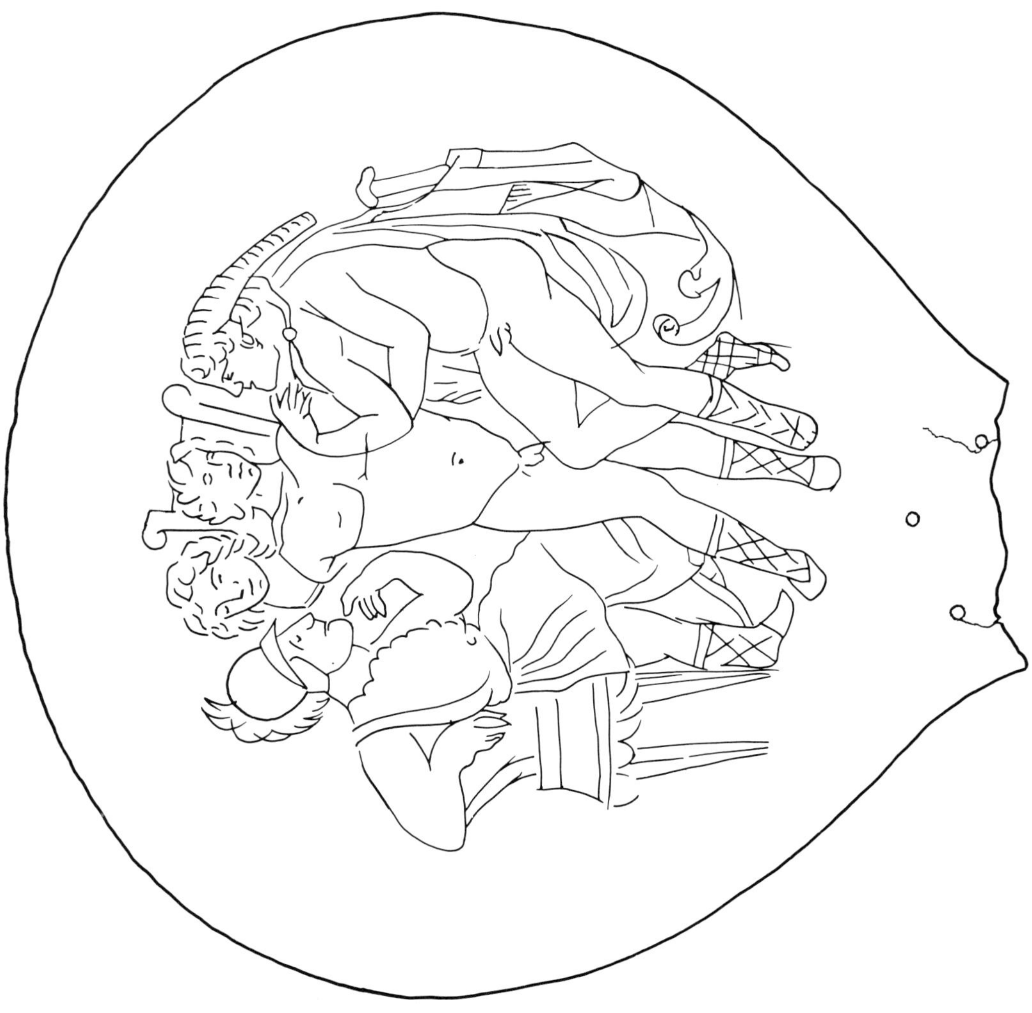

14a

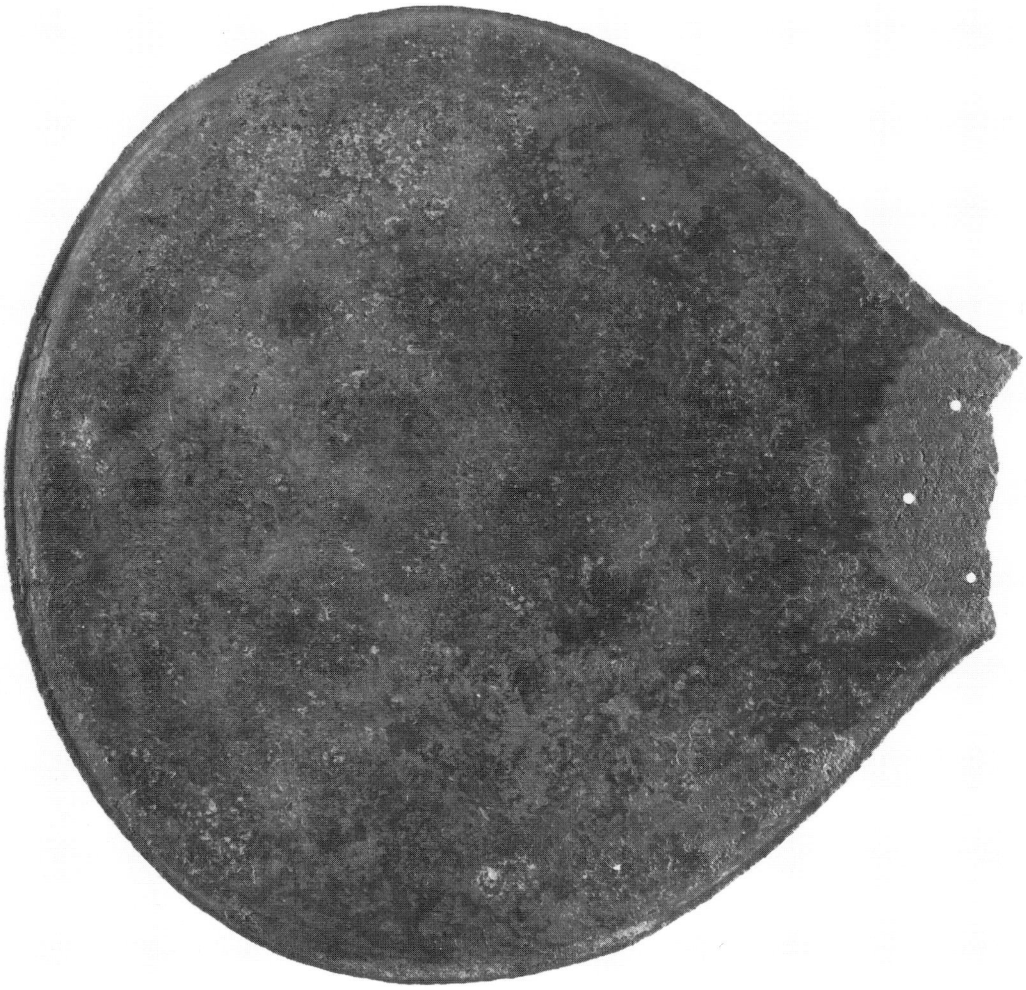

14d

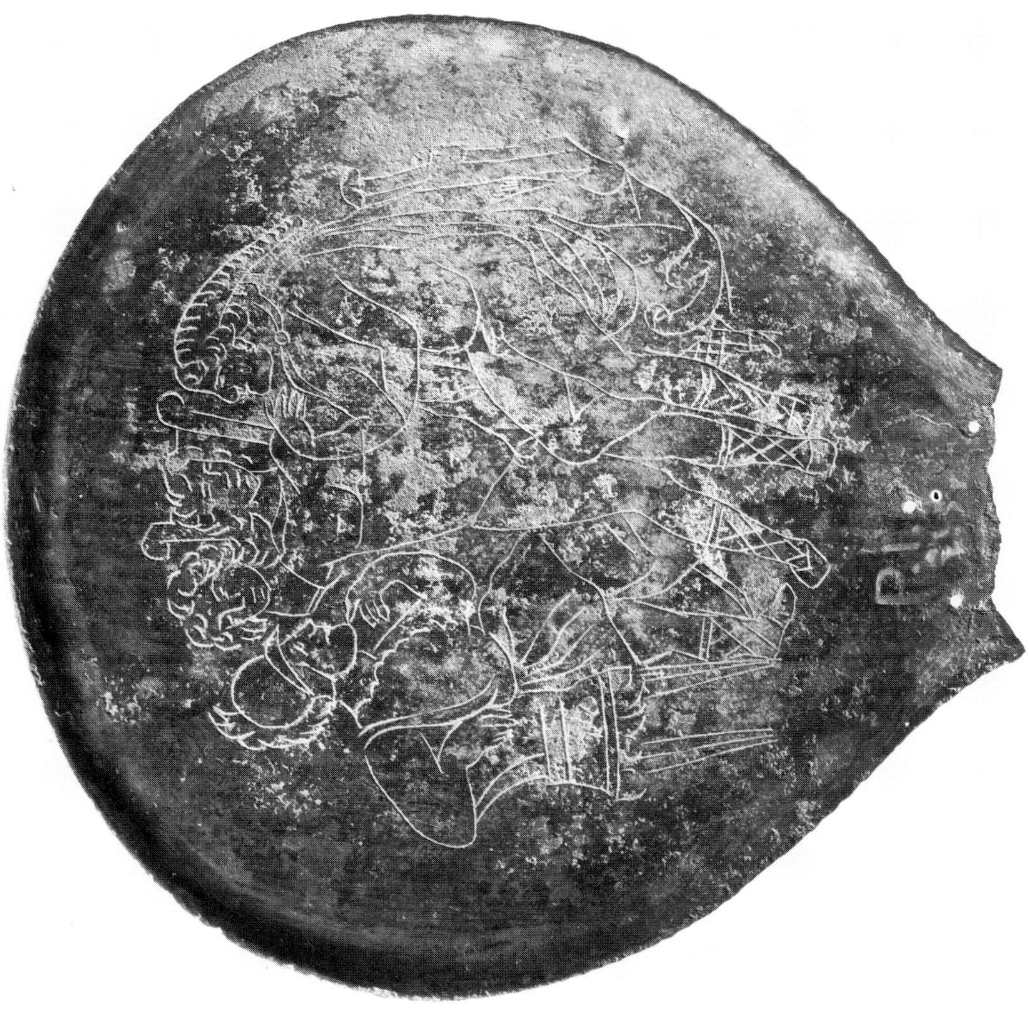

14b

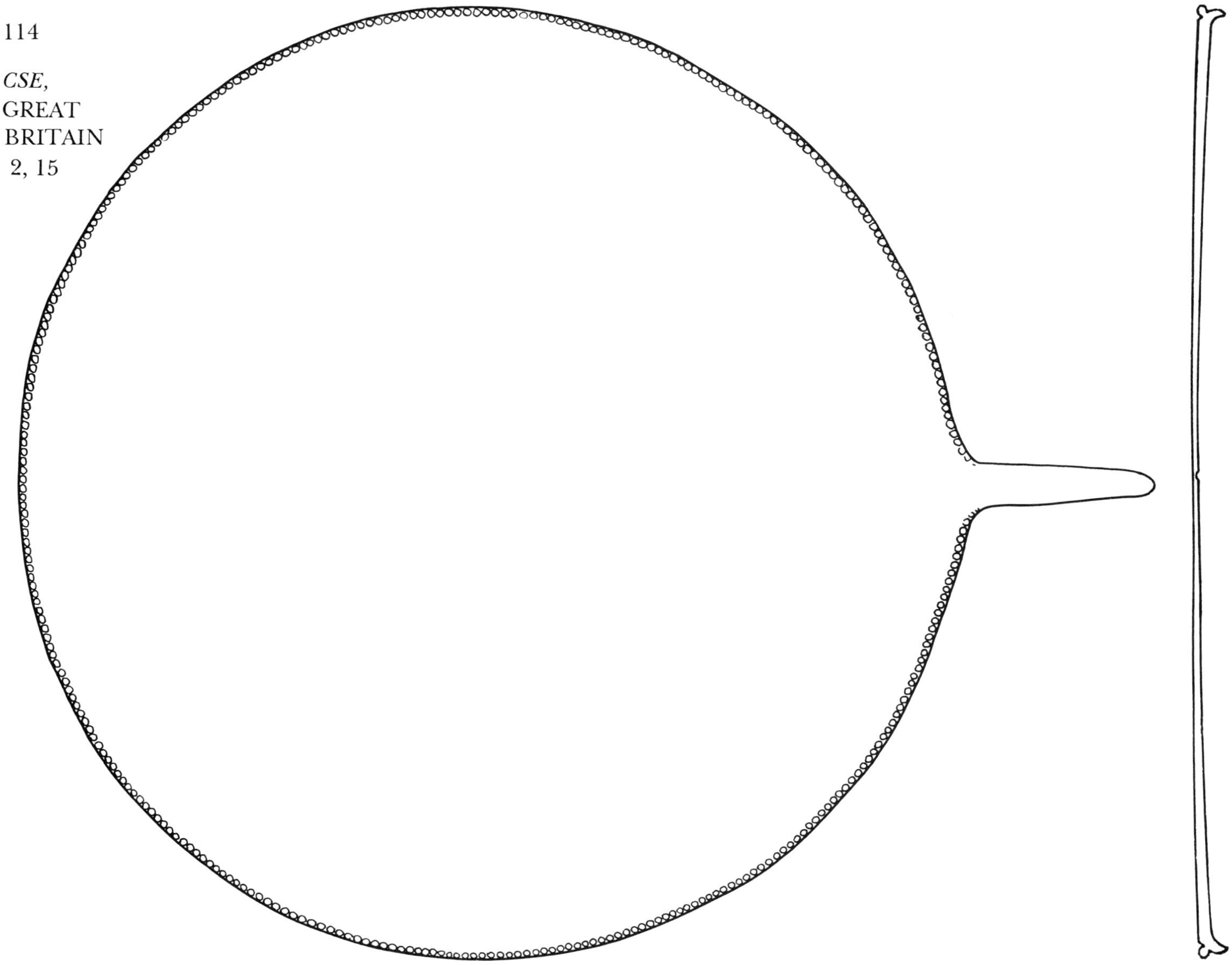

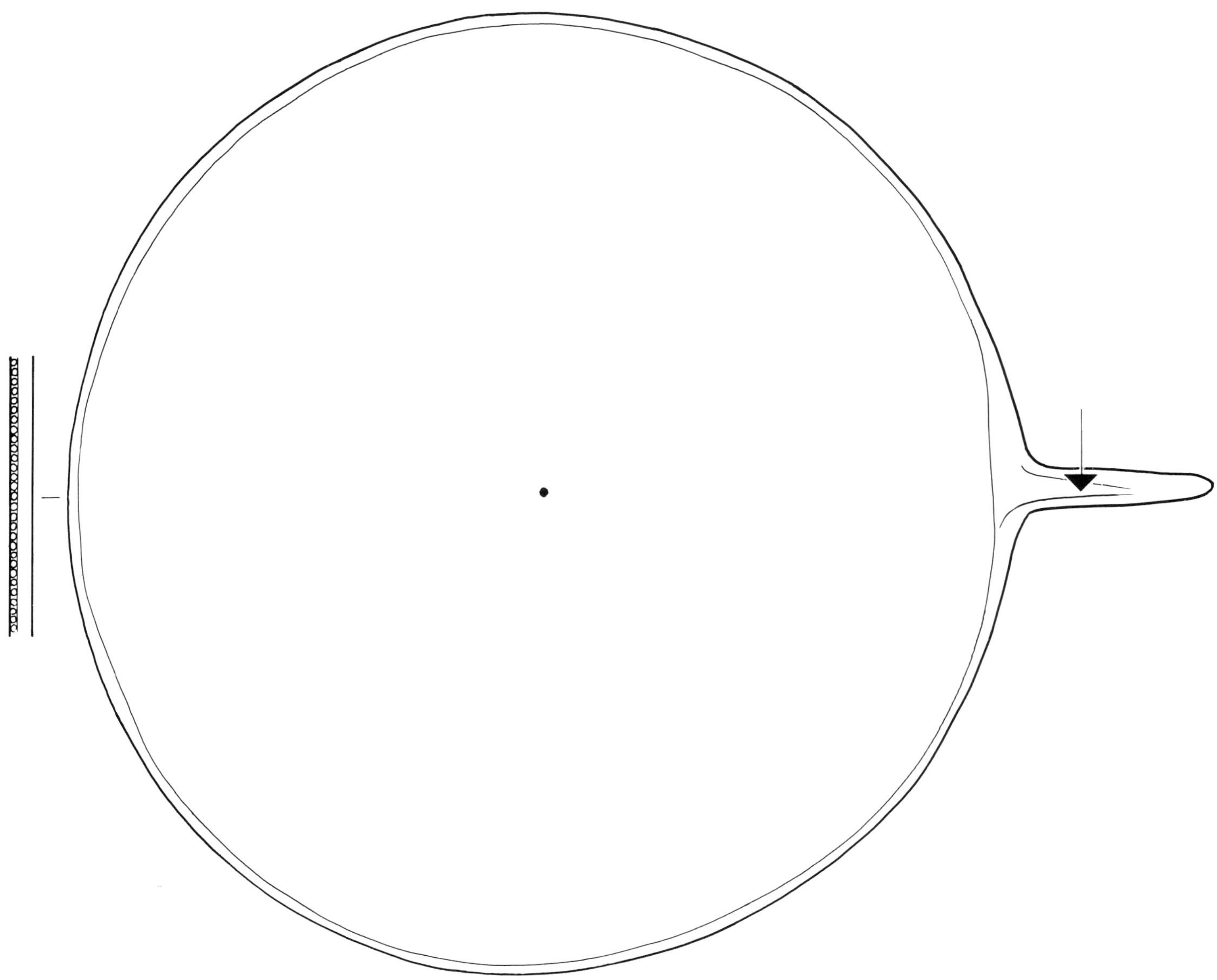

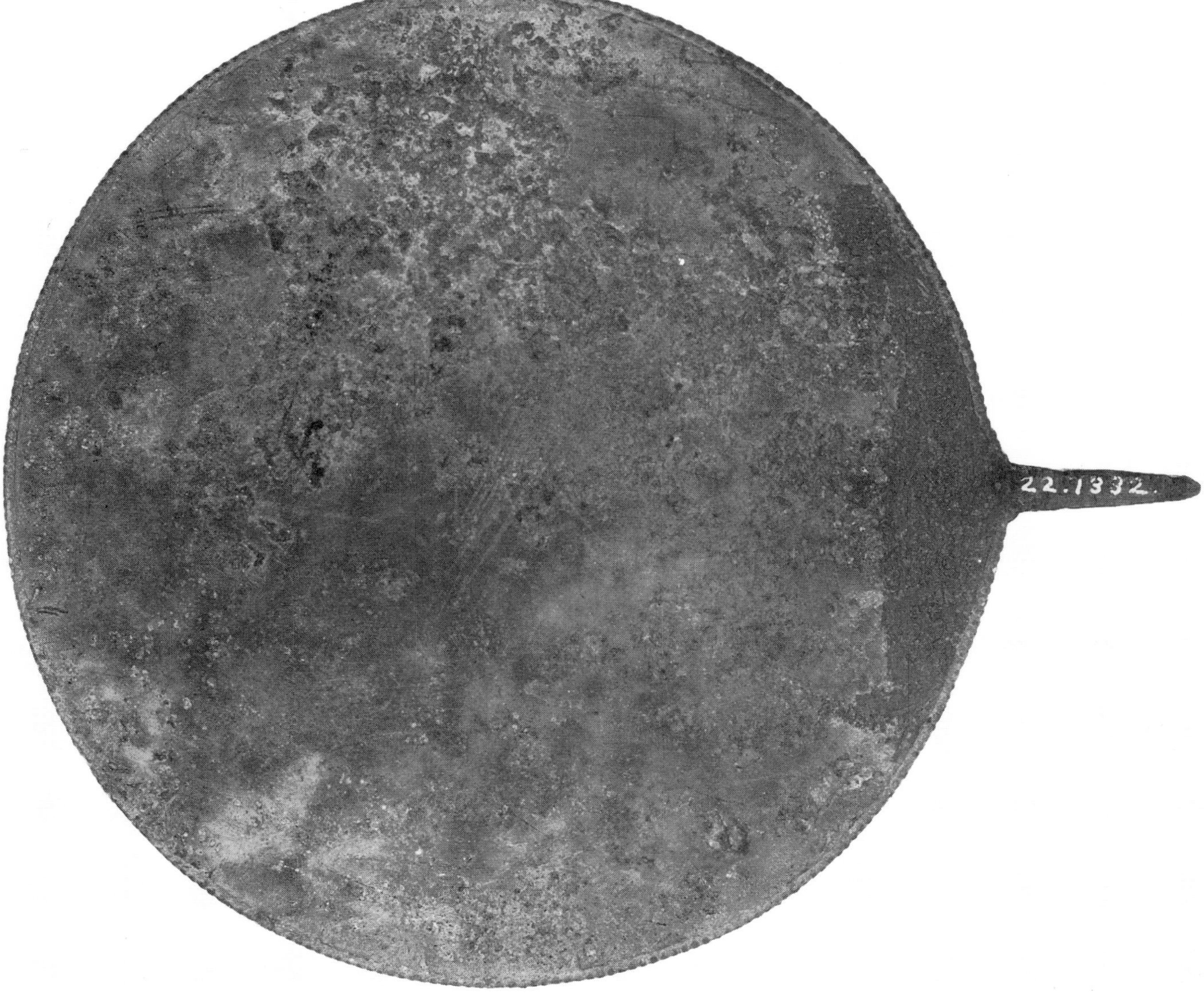

15d

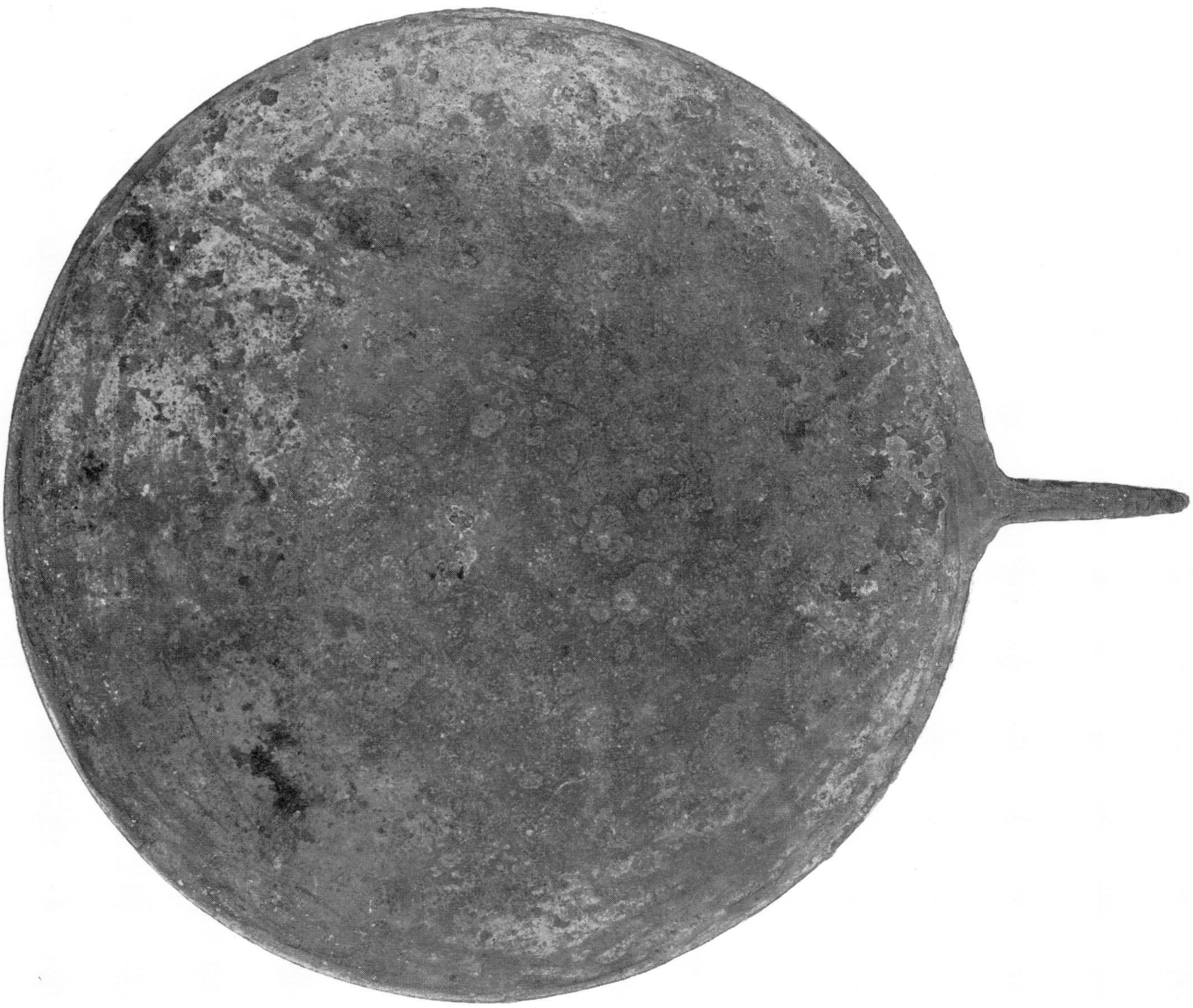

15b

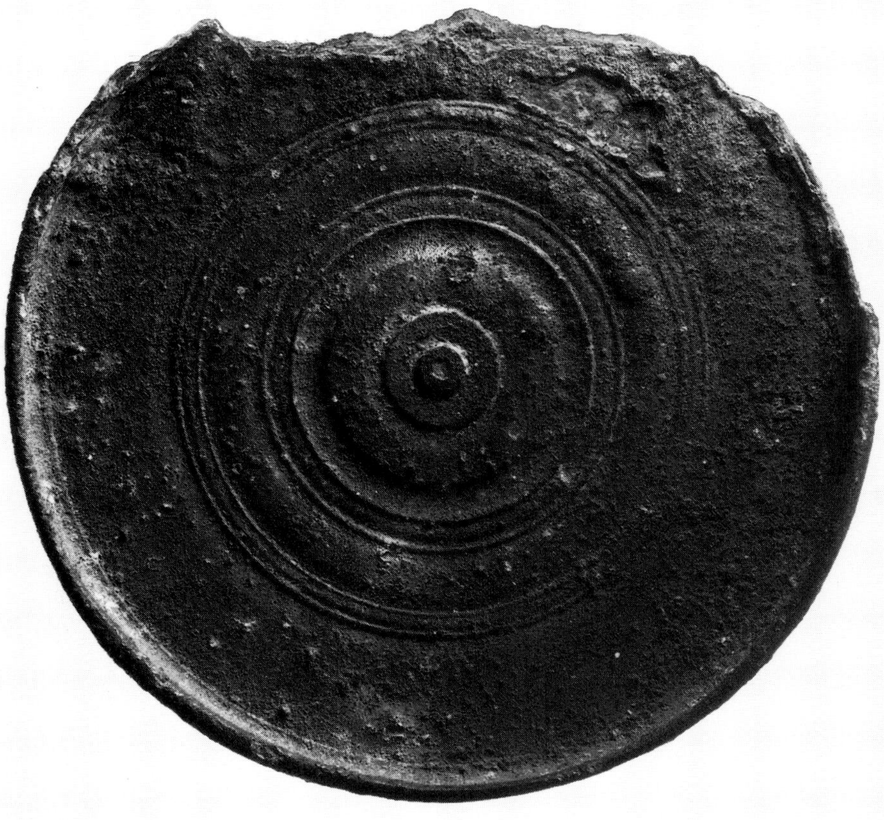

16b

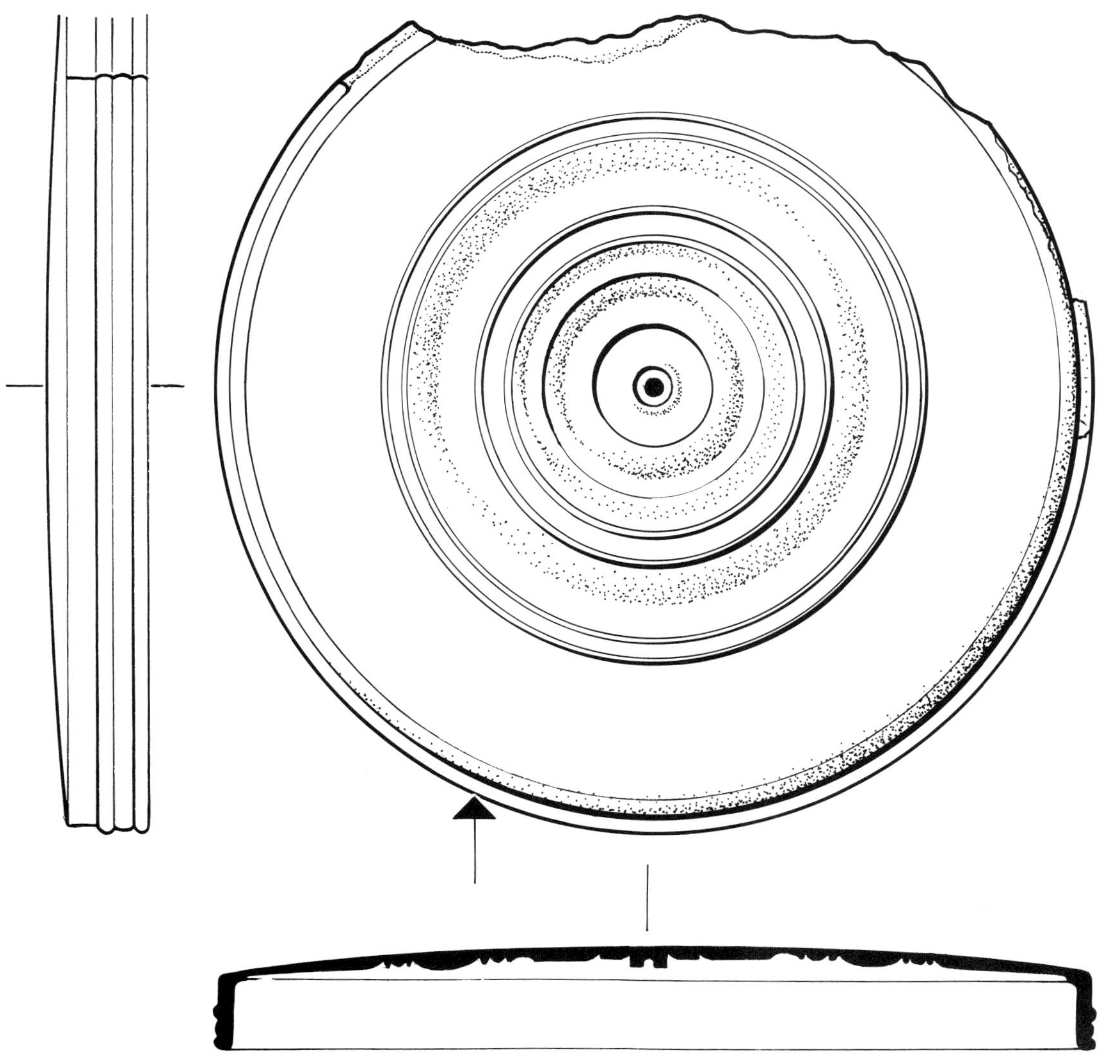

16a

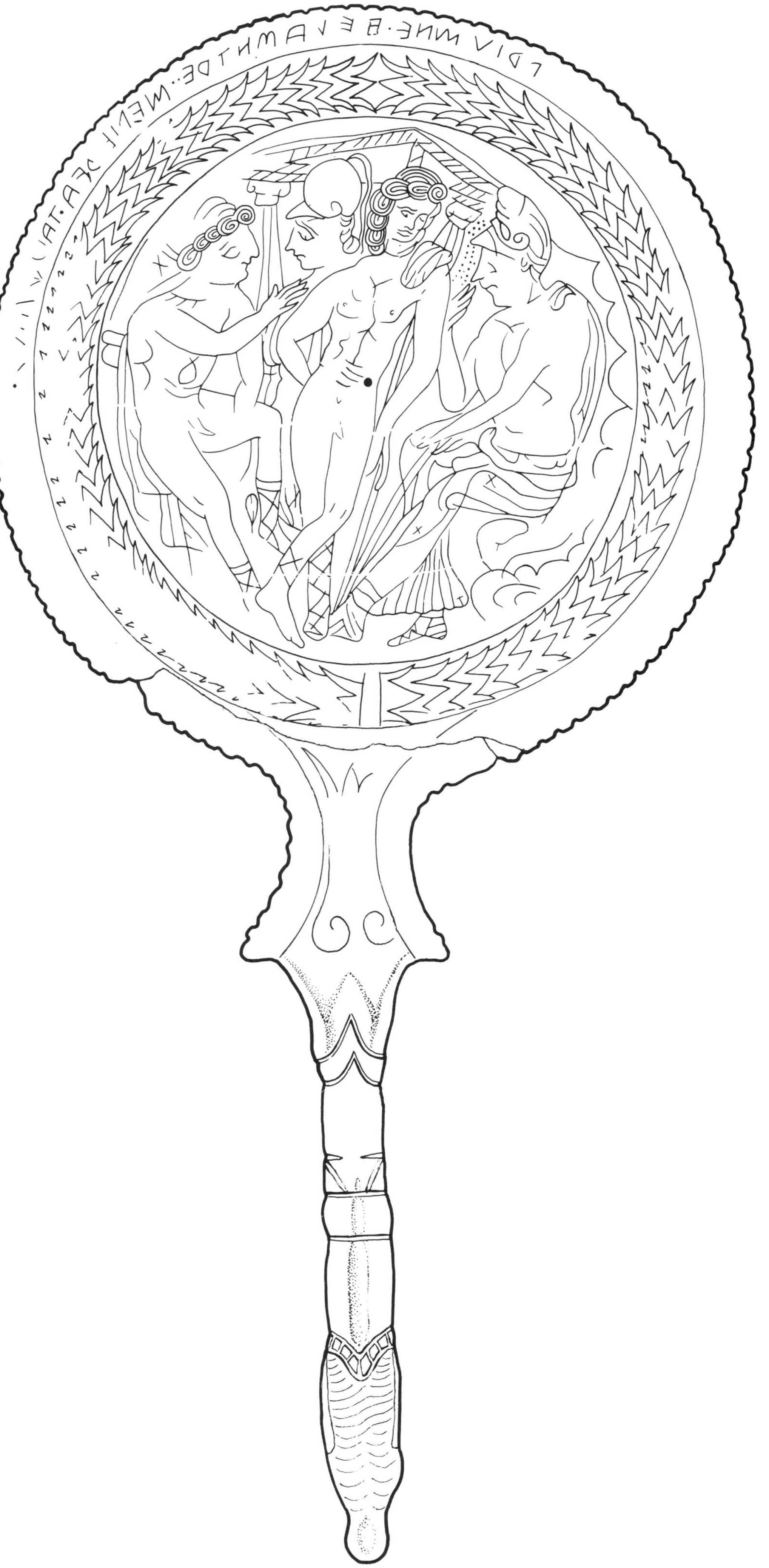

17a

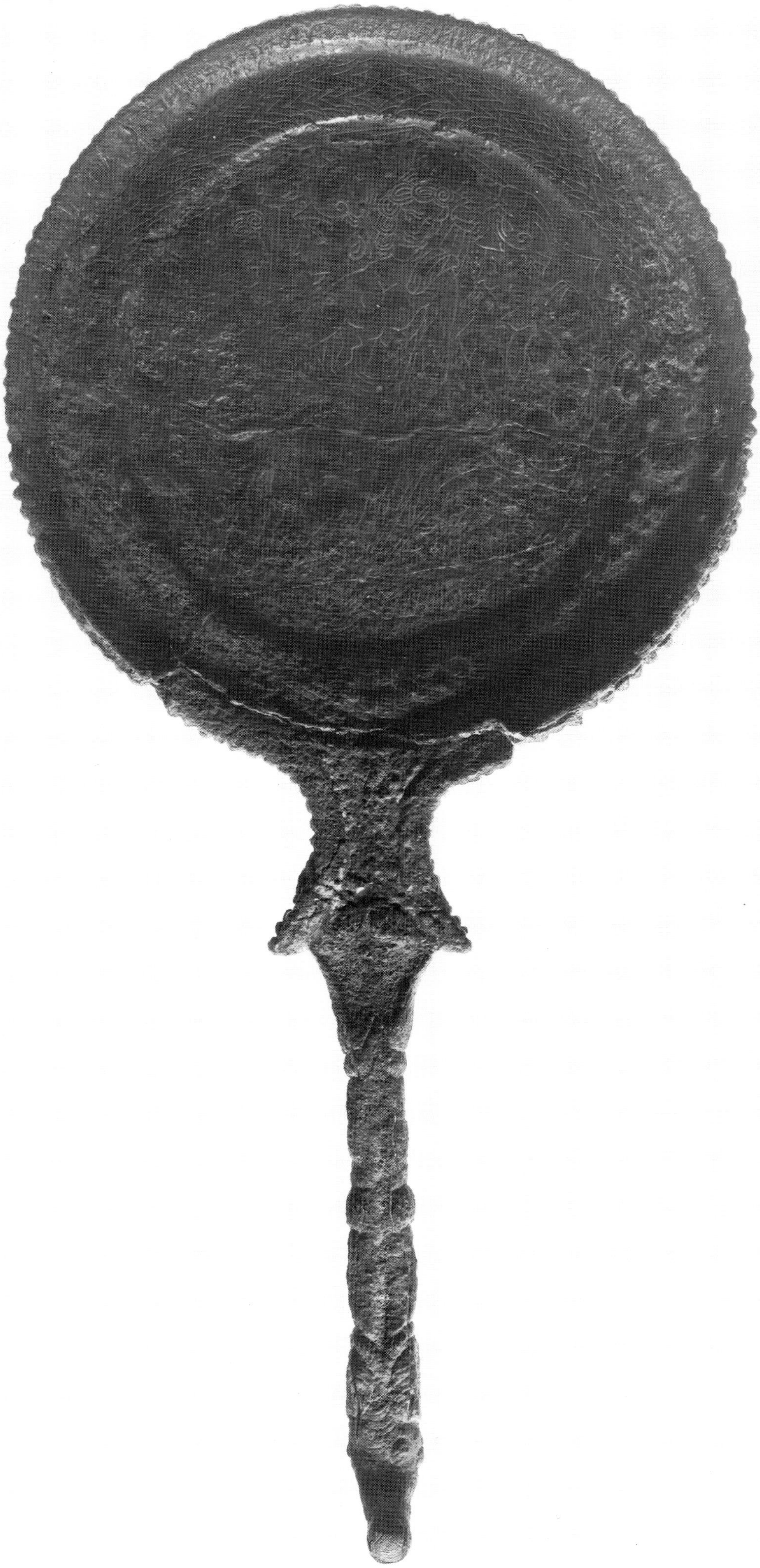

17b

CSE,
GREAT
BRITAIN
2, 17

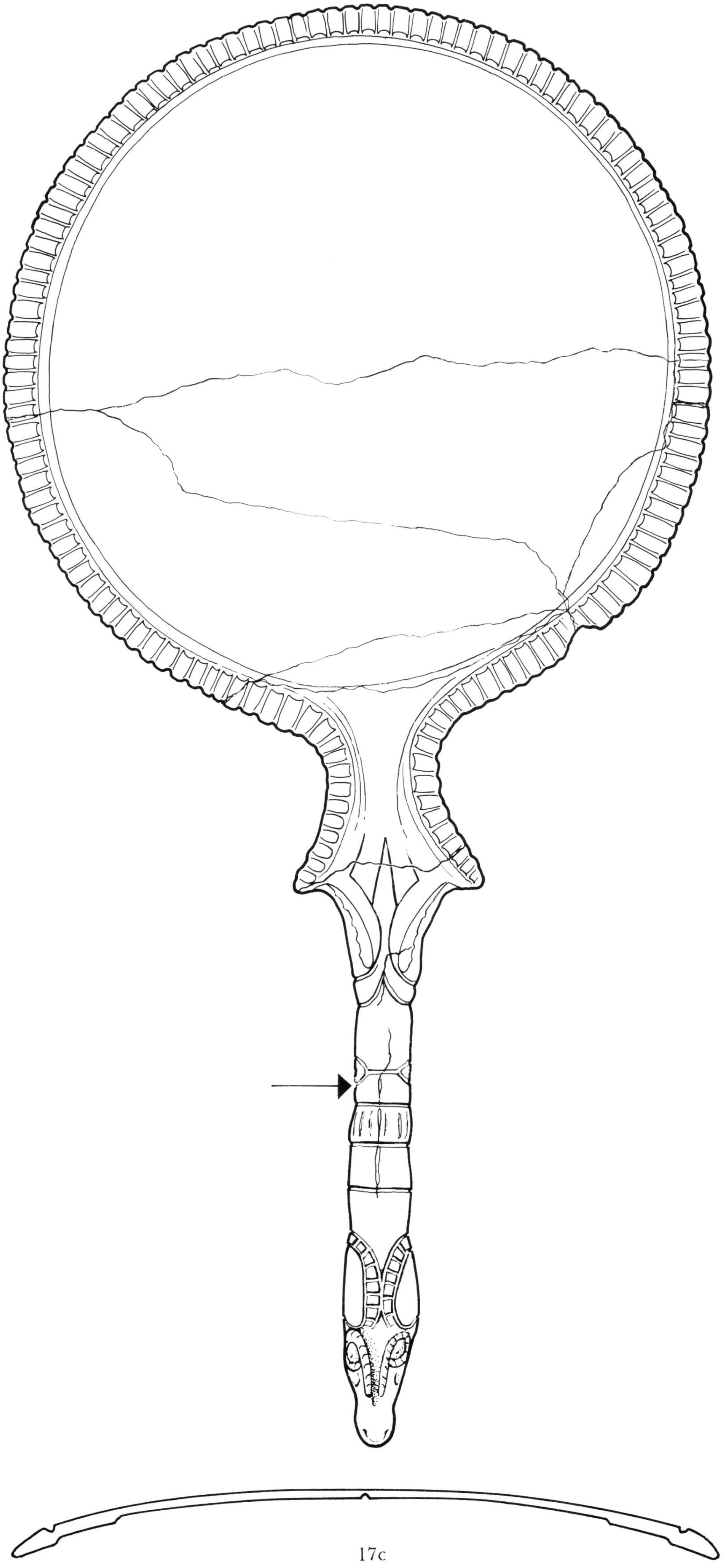

17c

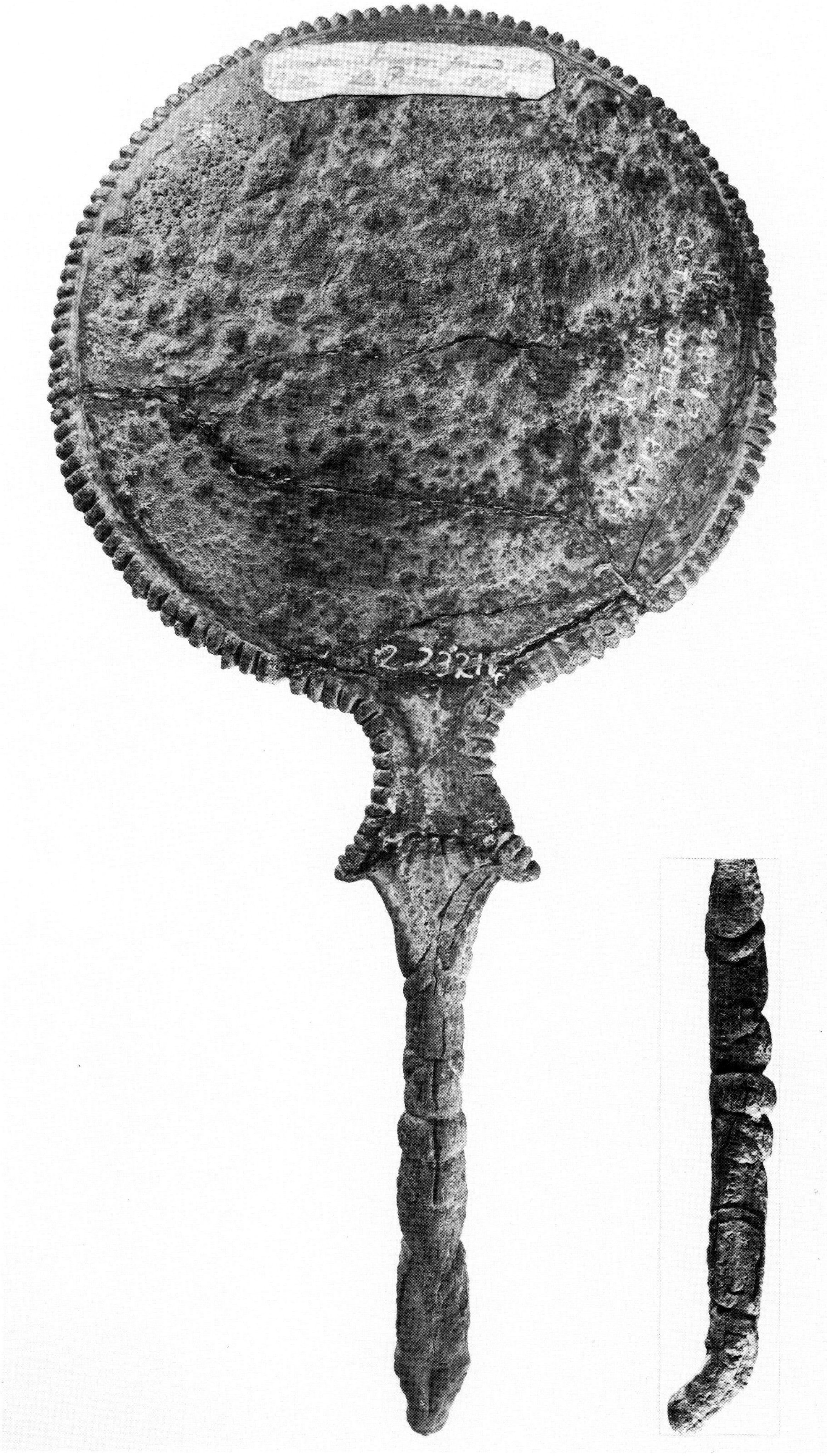

17d

CSE,
GREAT
BRITAIN
2, 18

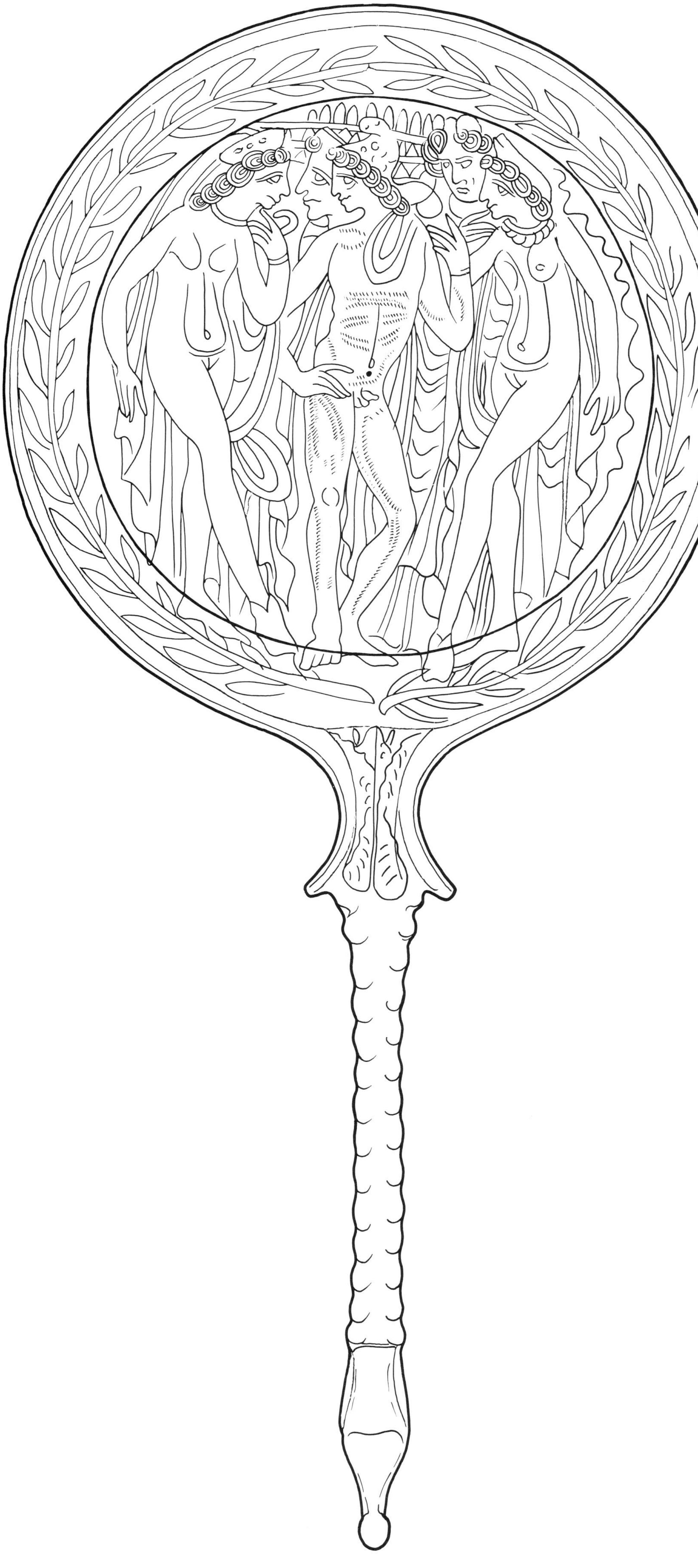

18a

CSE,
GREAT
BRITAIN
2, 18

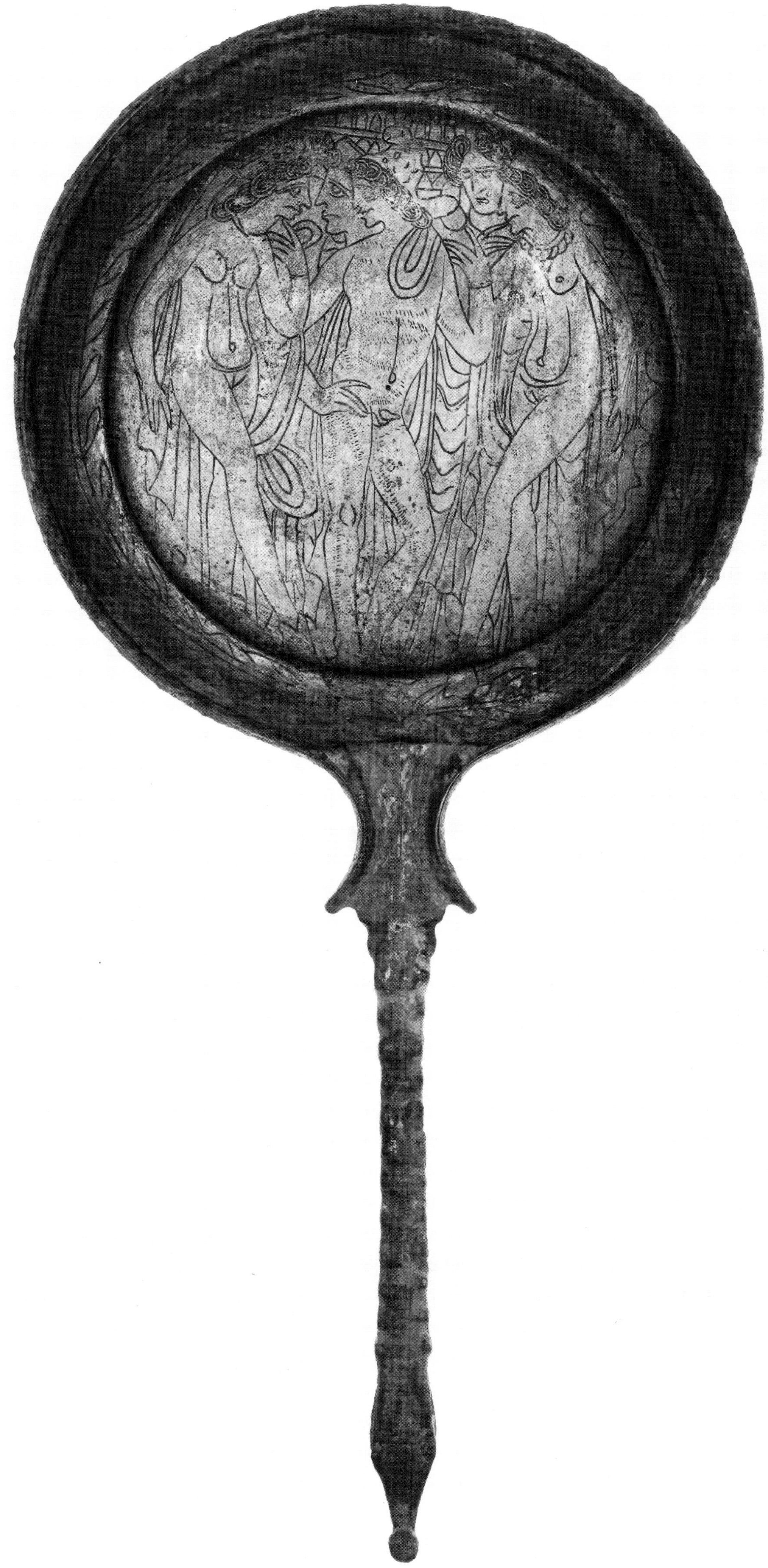

CSE,
GREAT
BRITAIN
2, 18

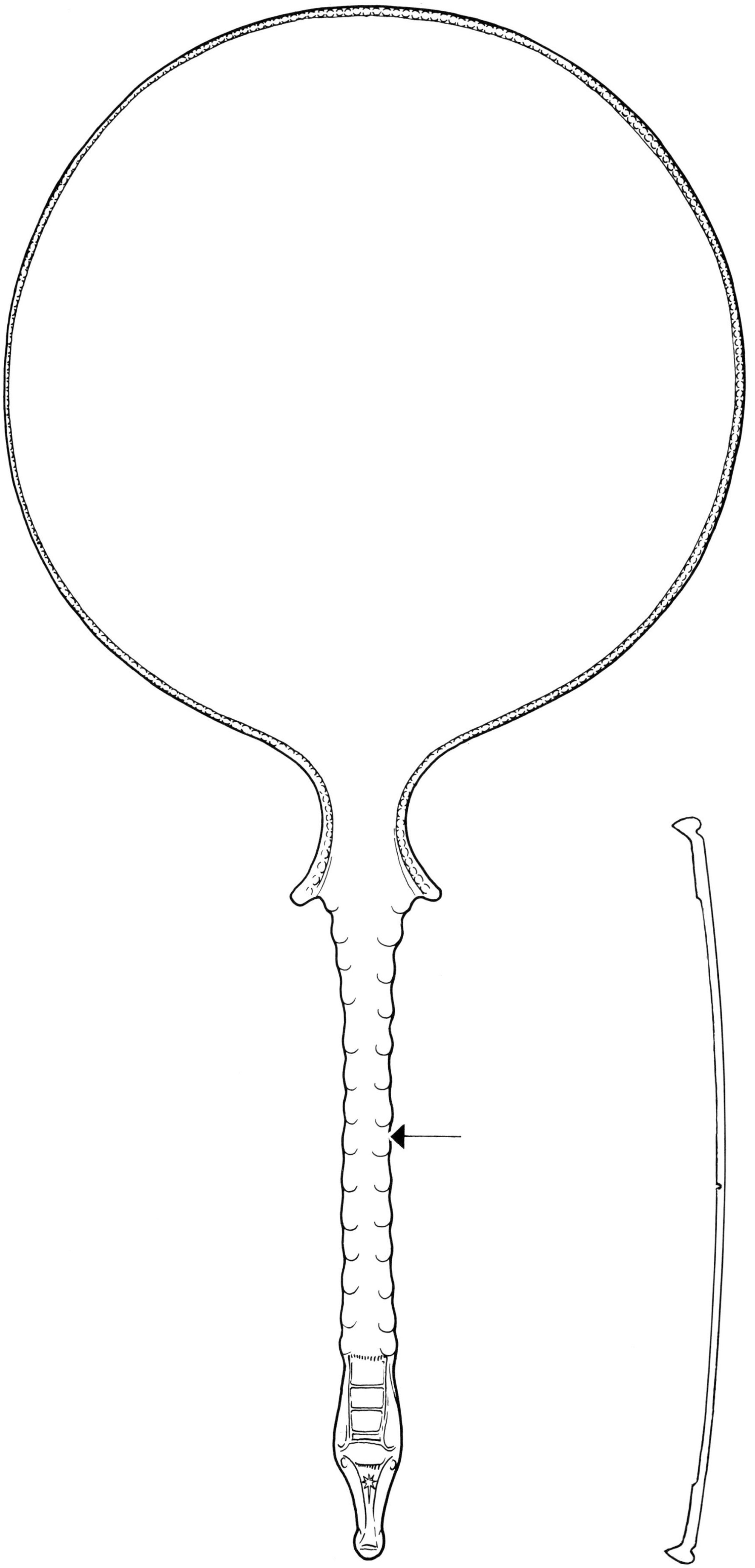

18c

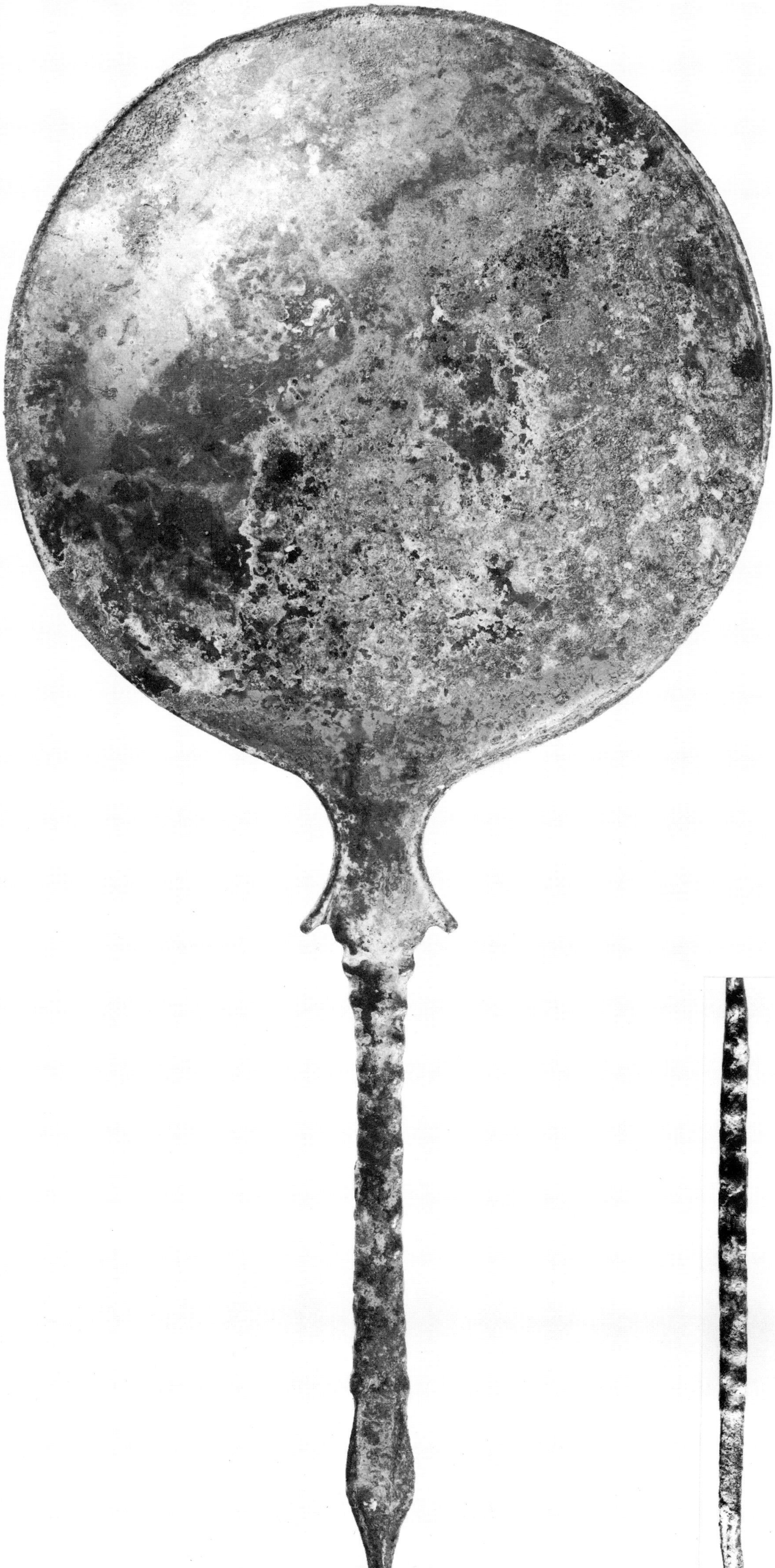

18d

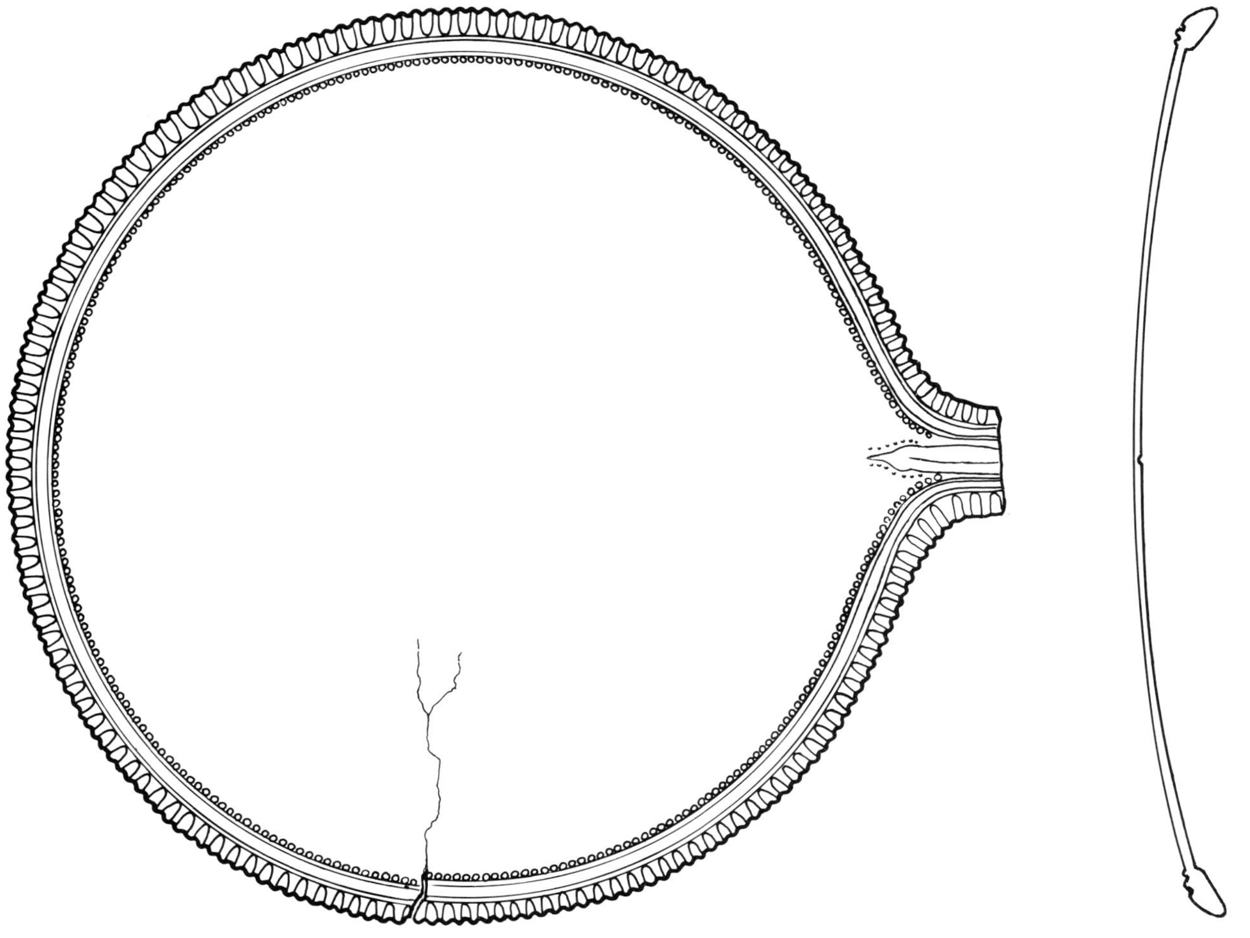

19c

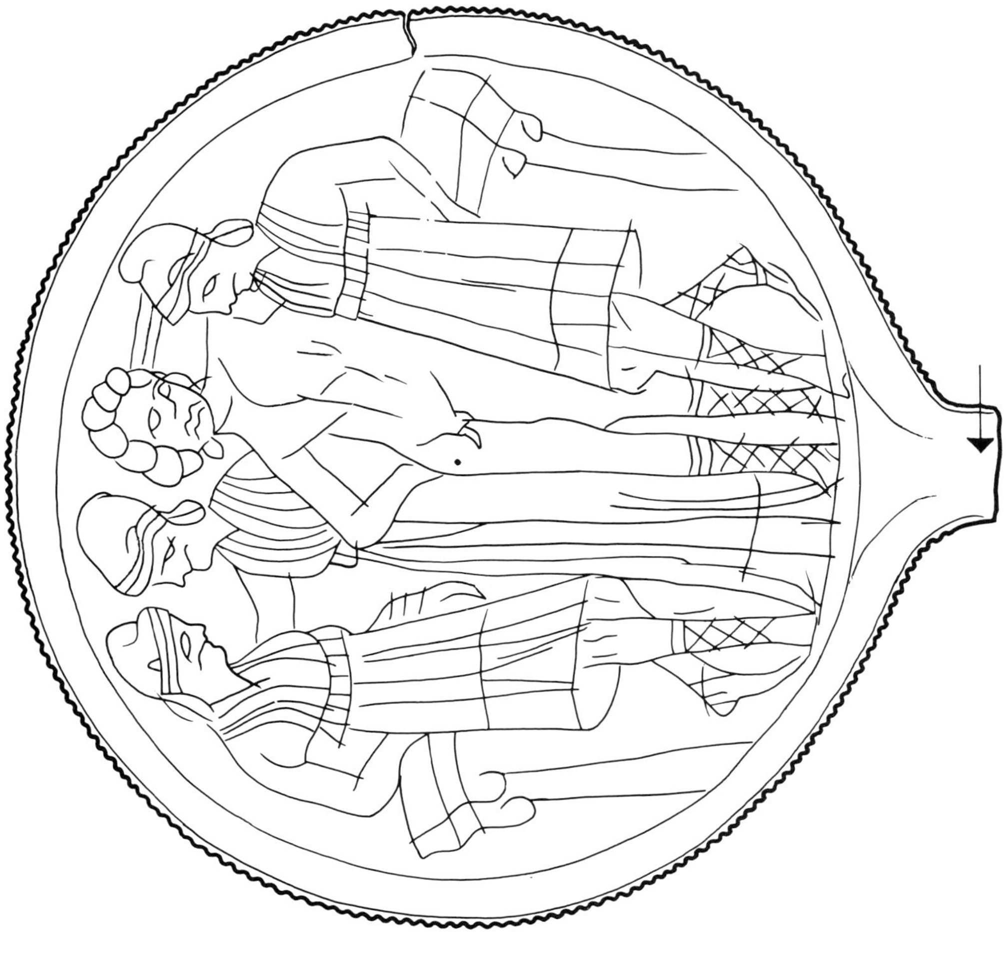

19a

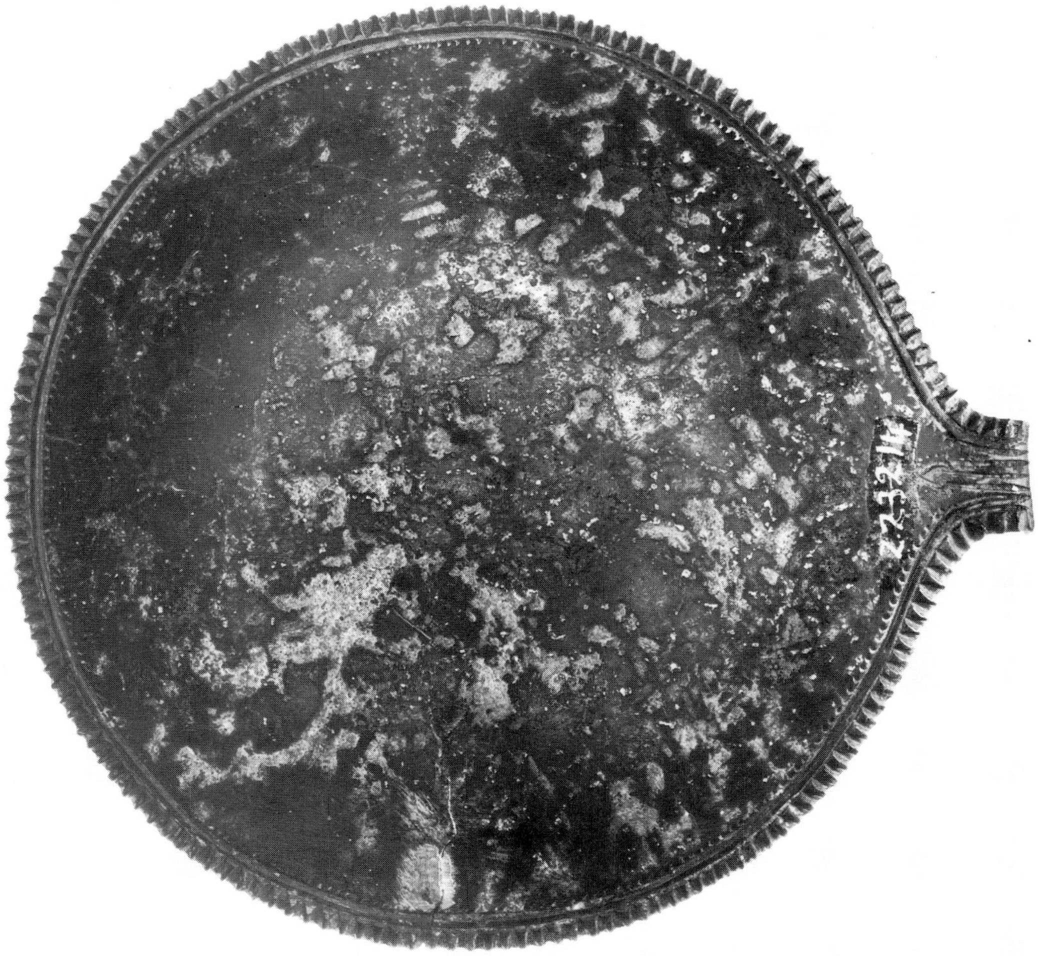

19d

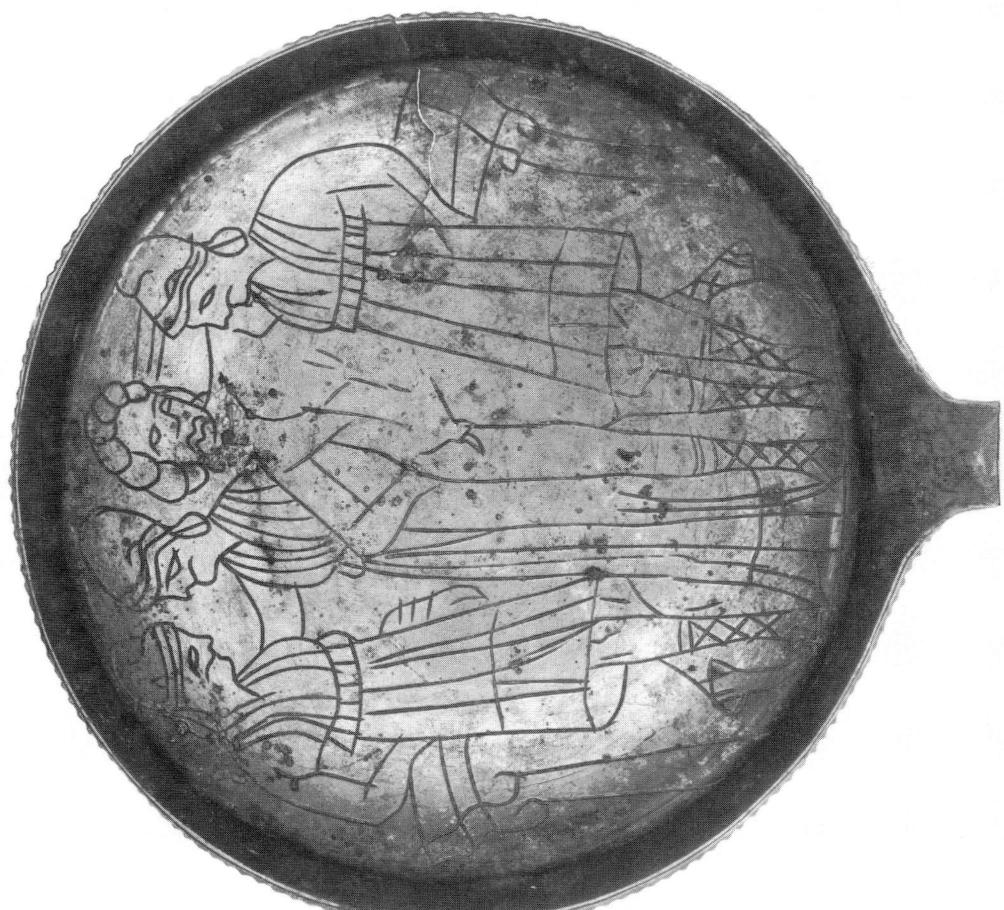

19b

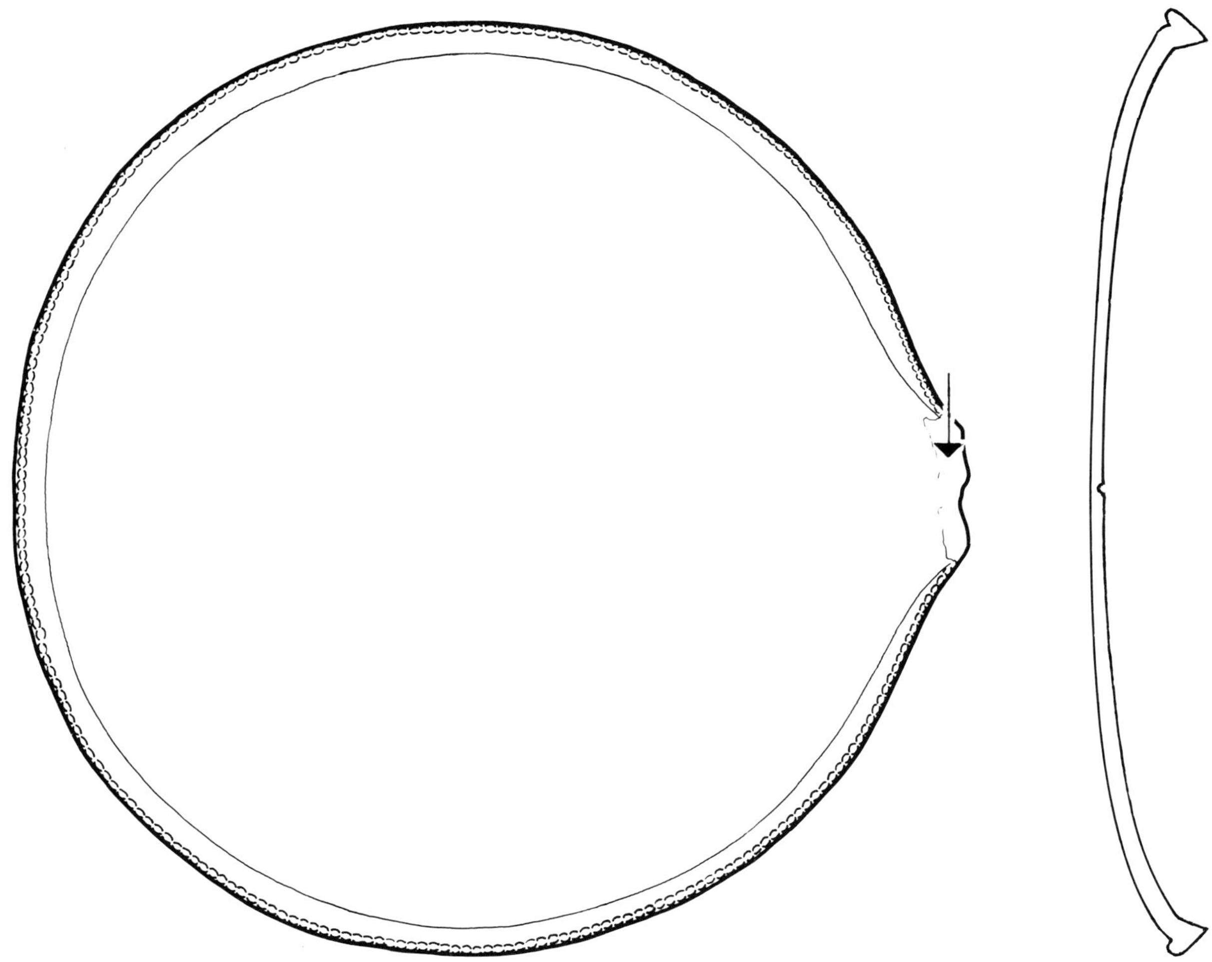

20c

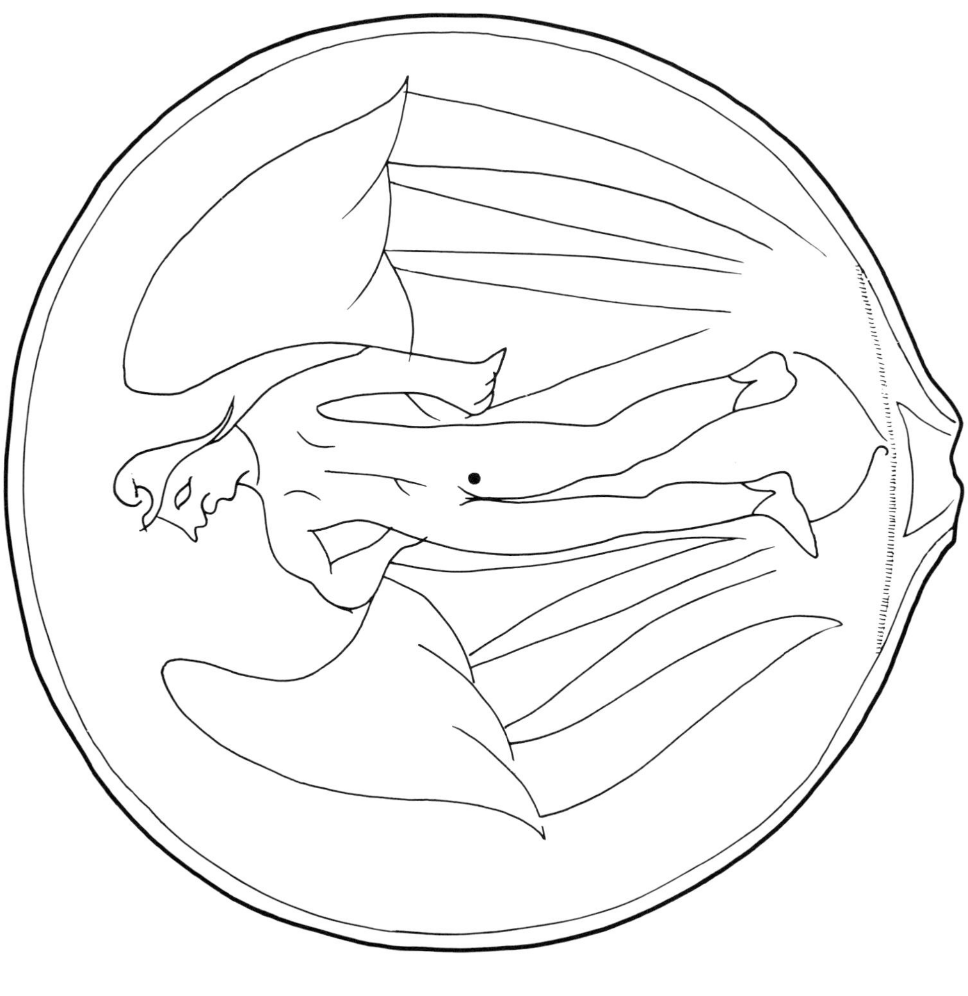

20a

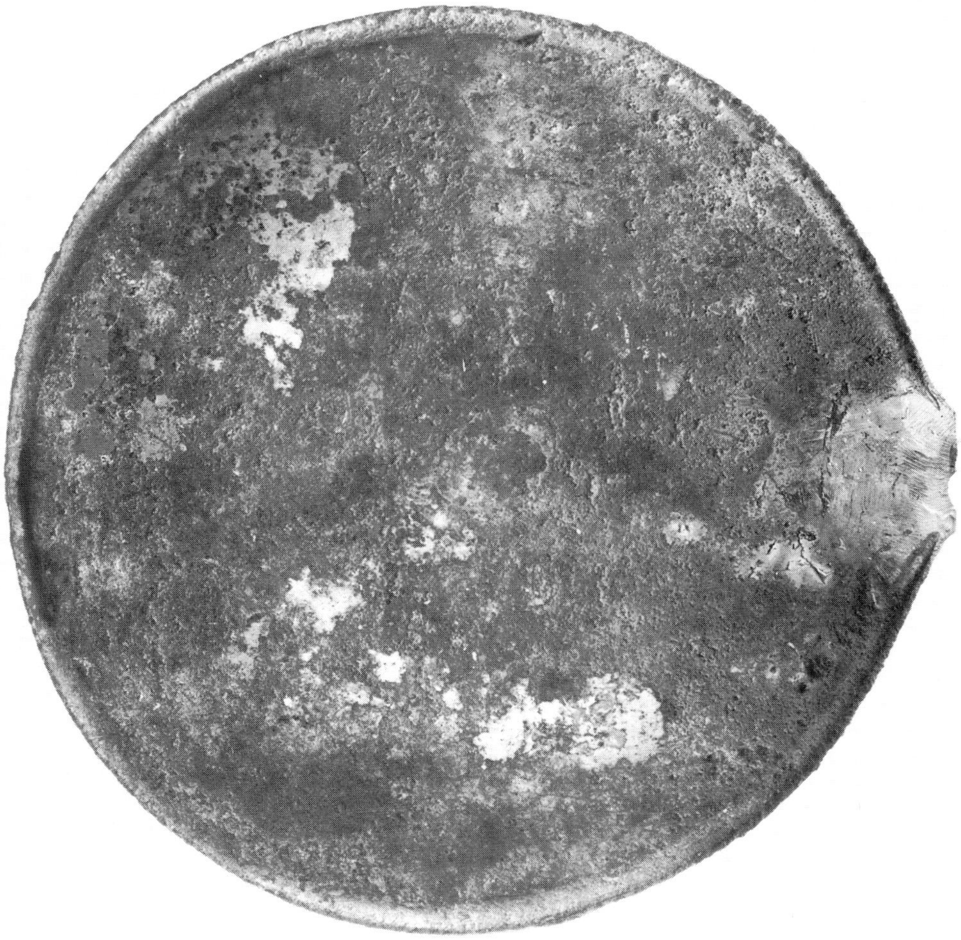

20d

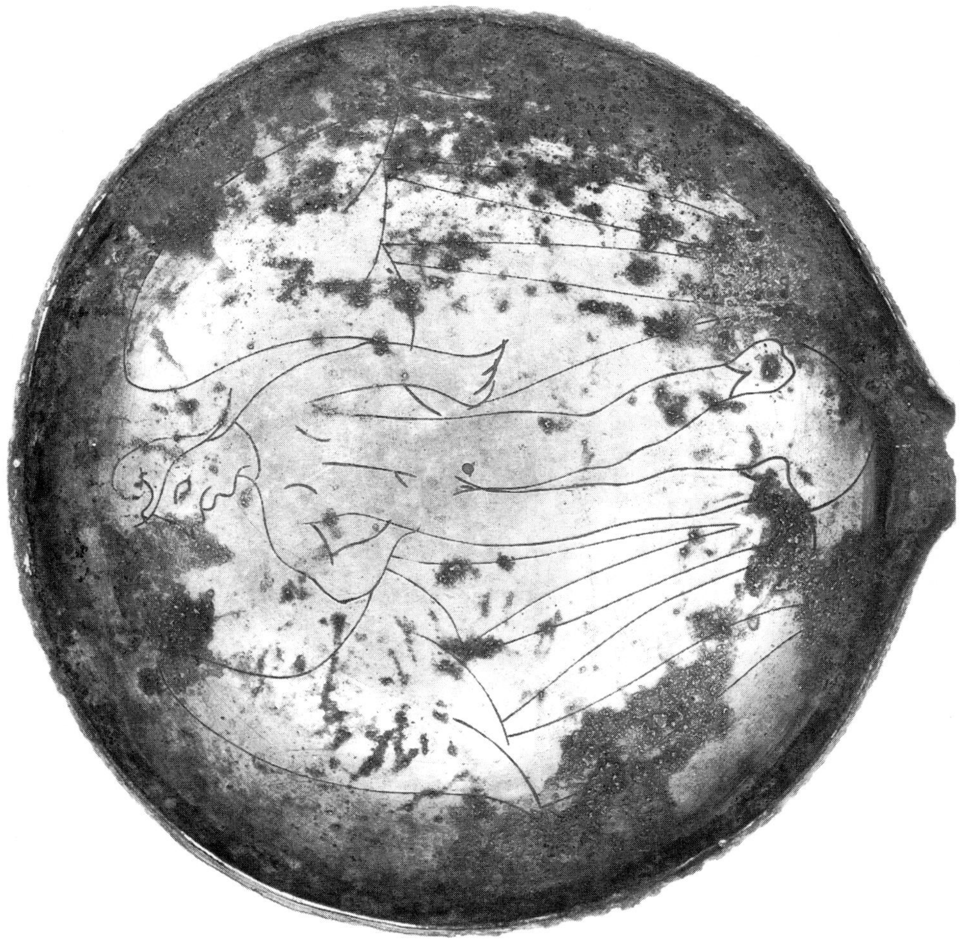

20b

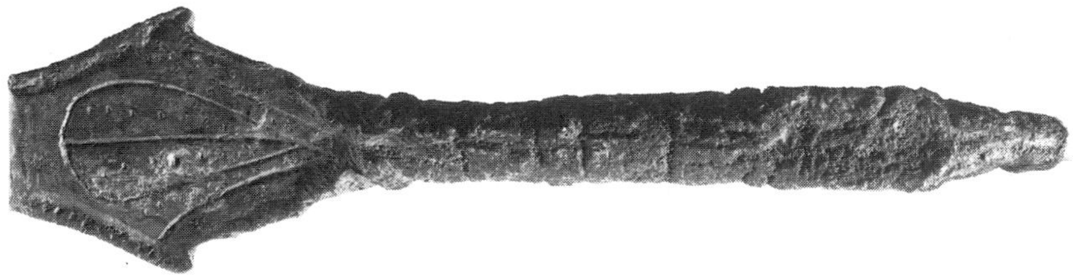
22d

22e

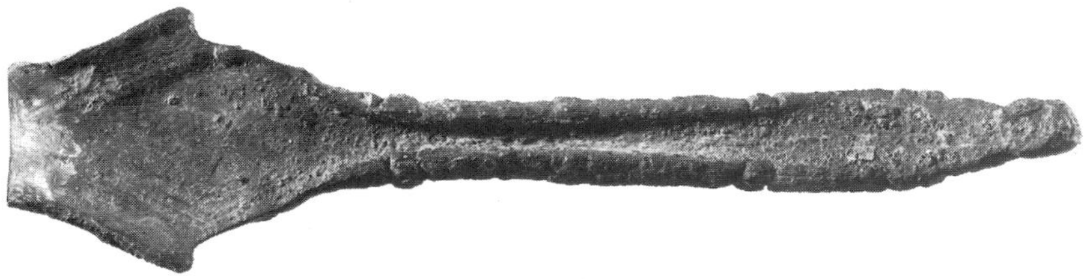
22b

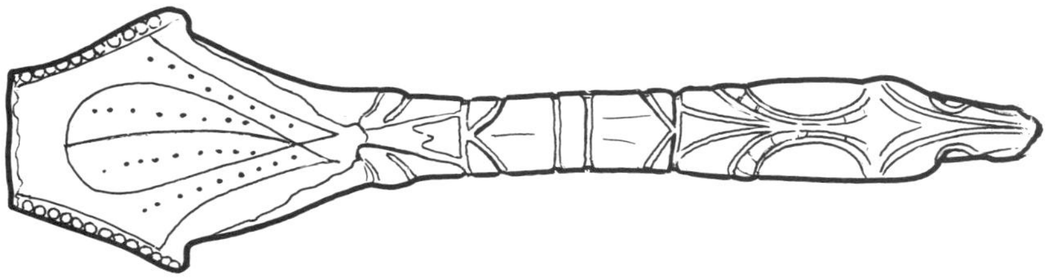
22c

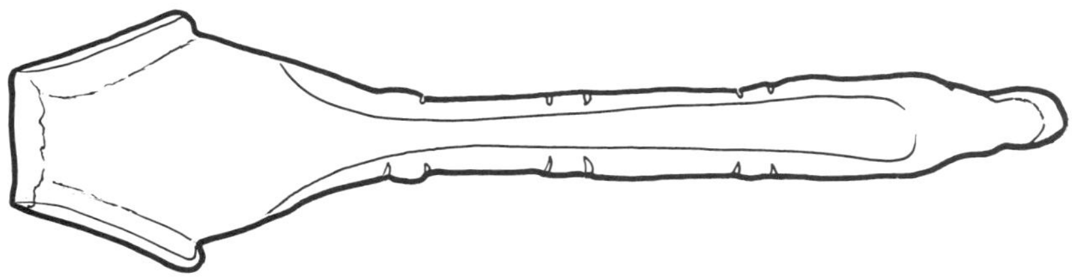
22a

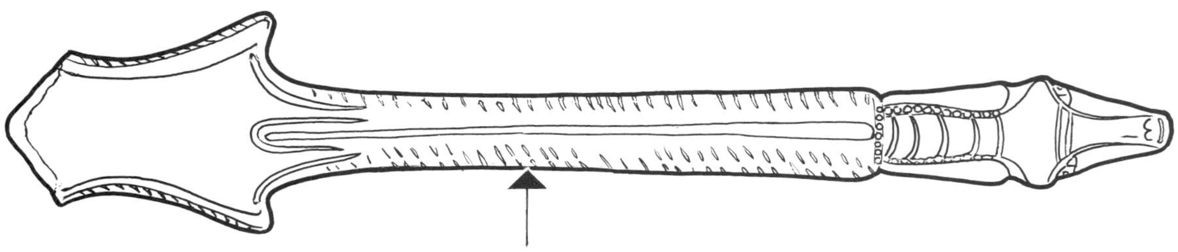
21c

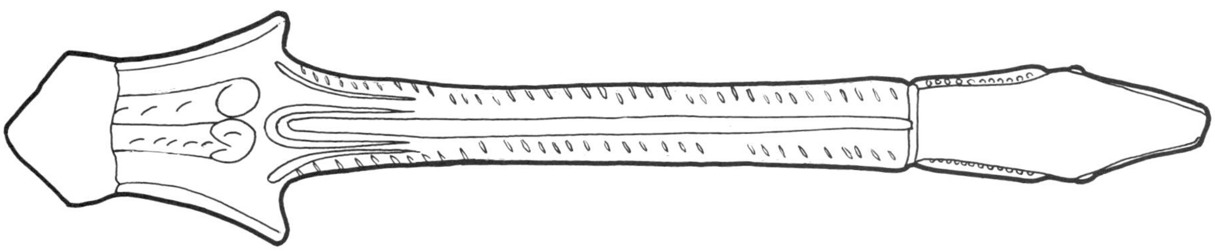
21a

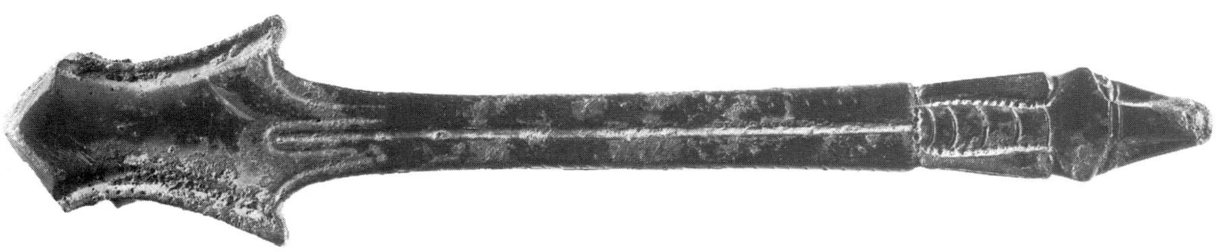
21d

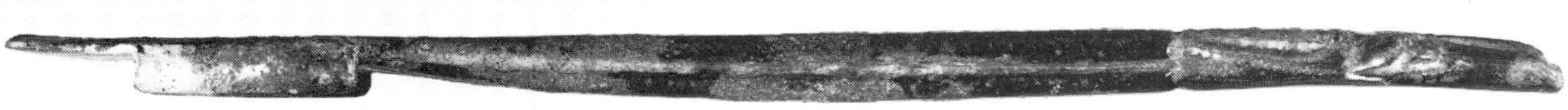
21e

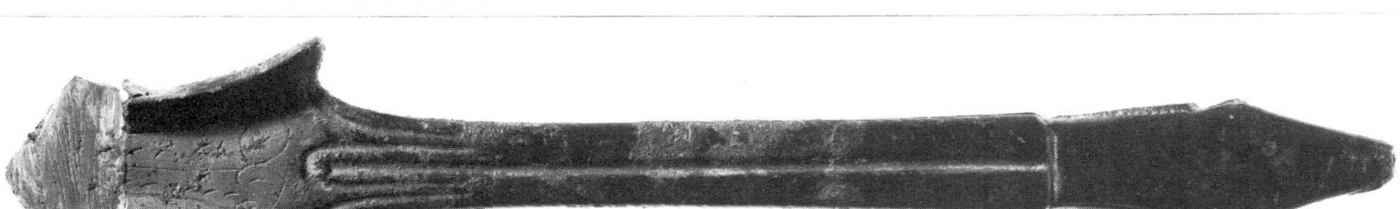
21b

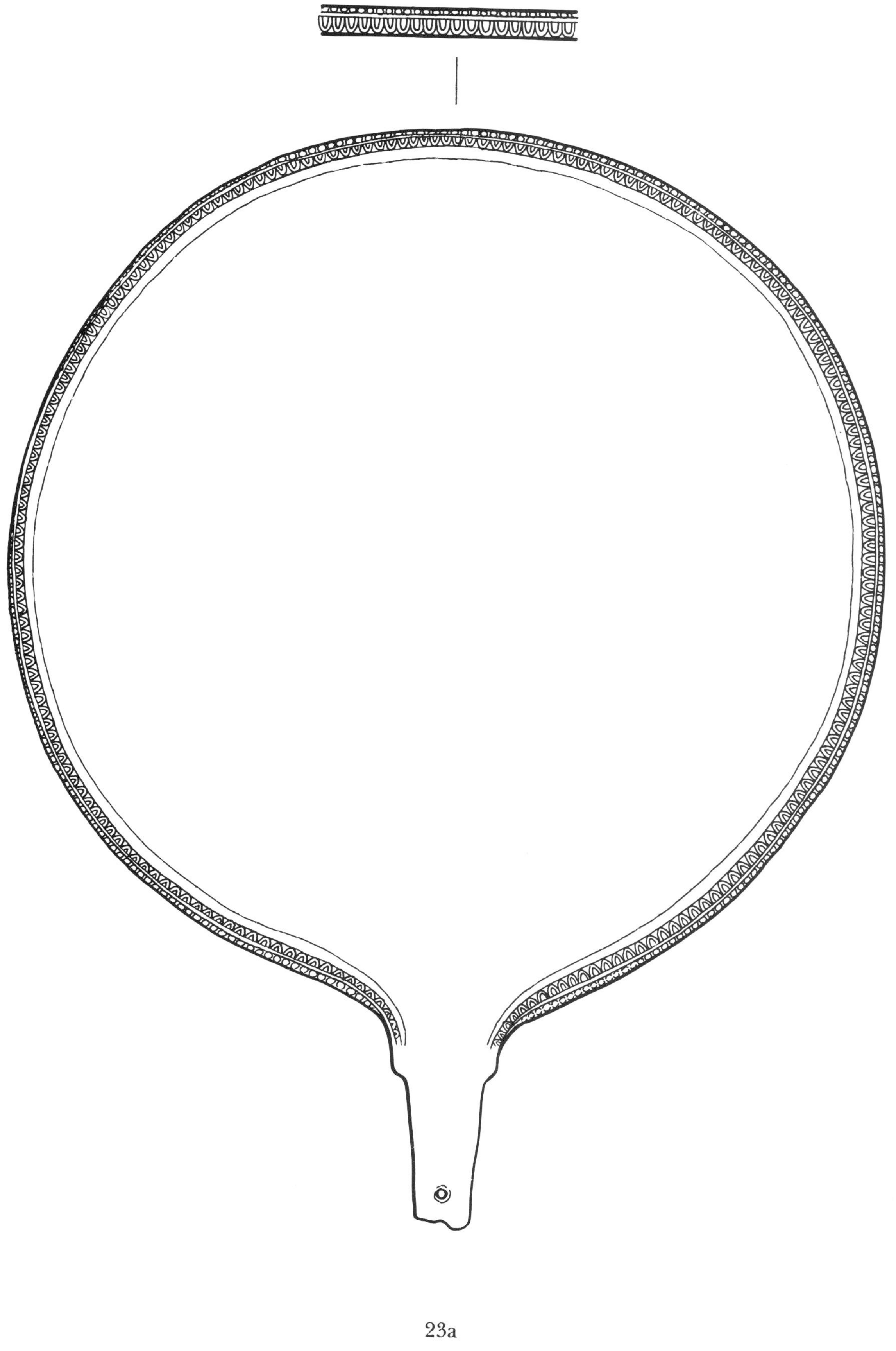

23a

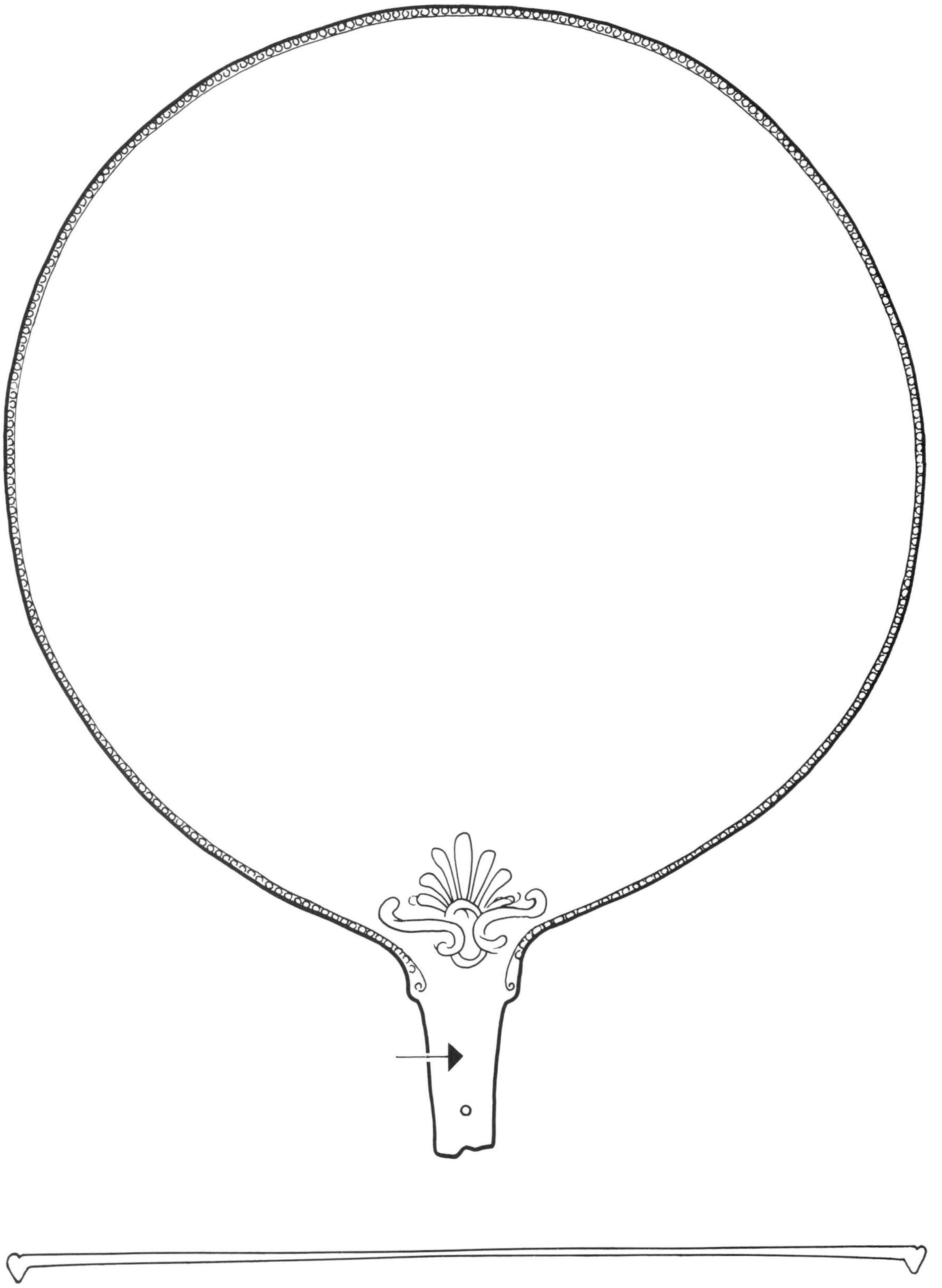

23b

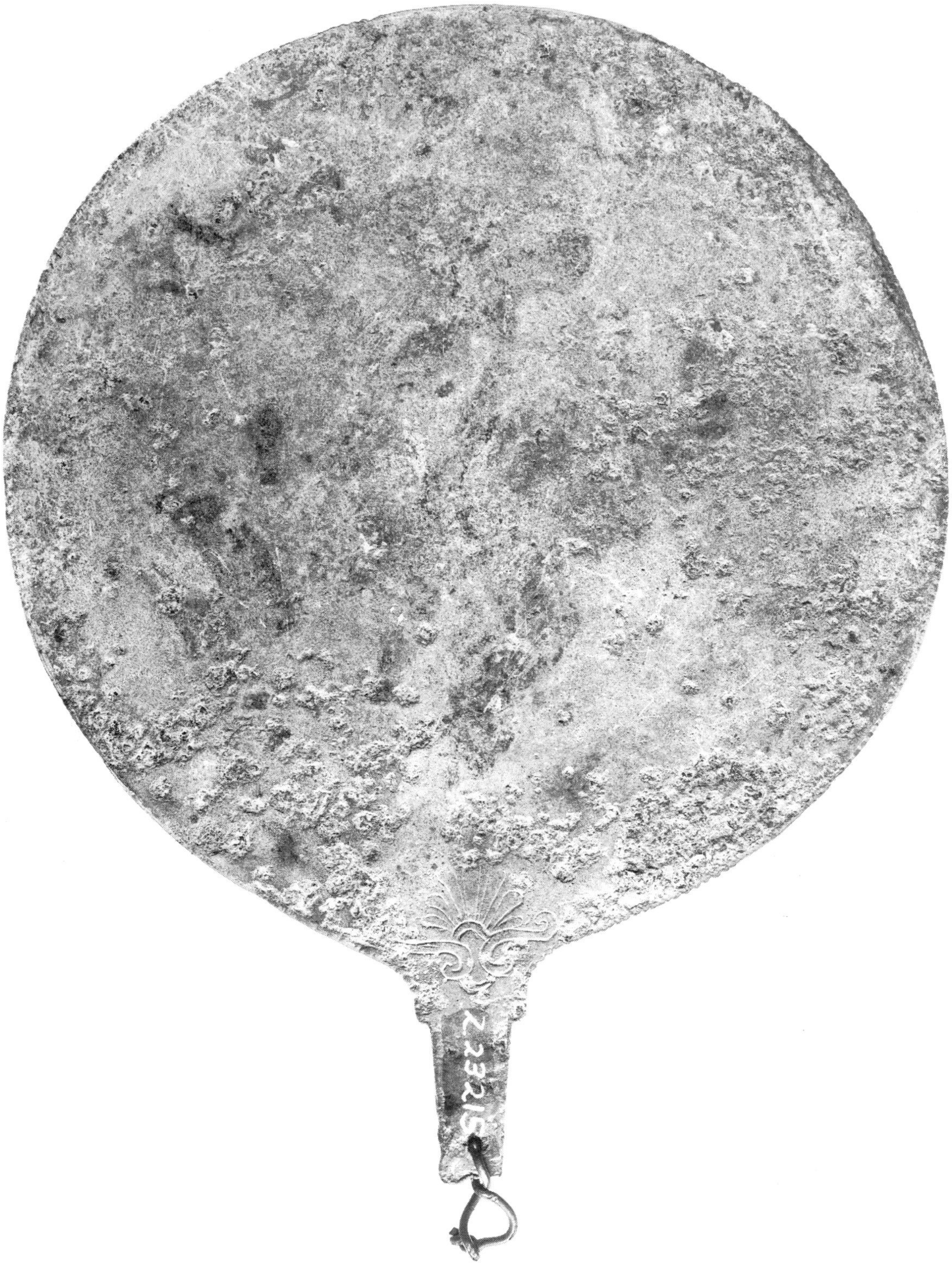

23c

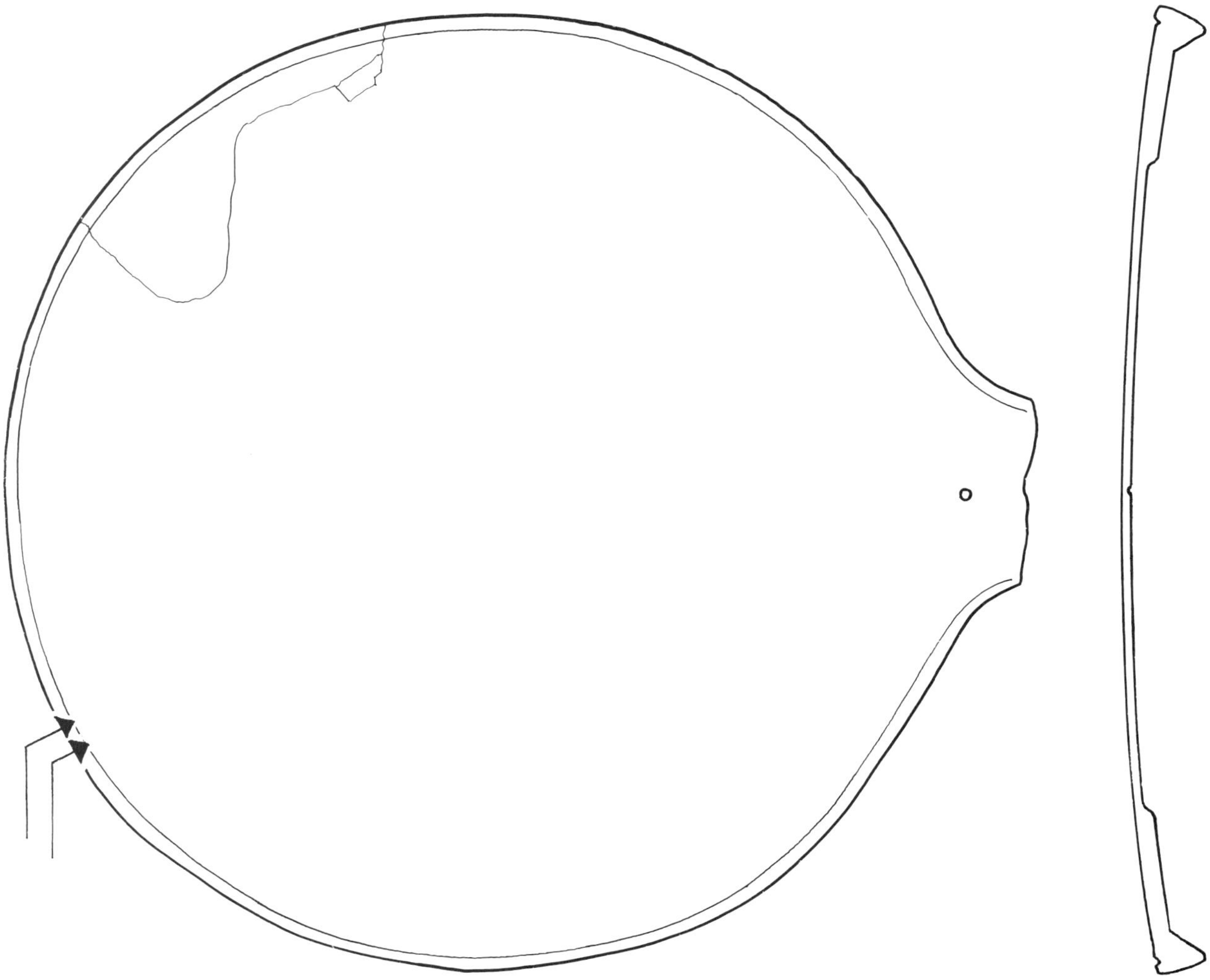

24c

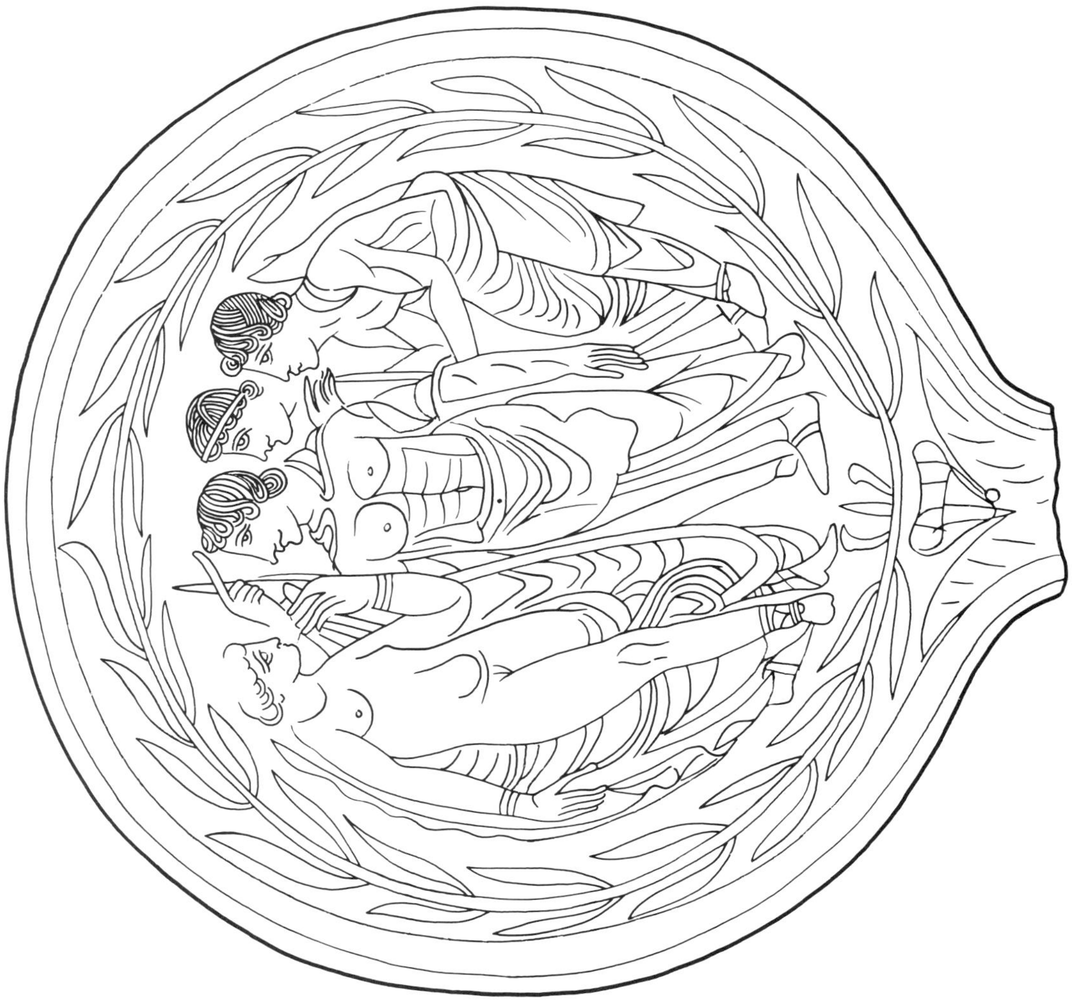

24a

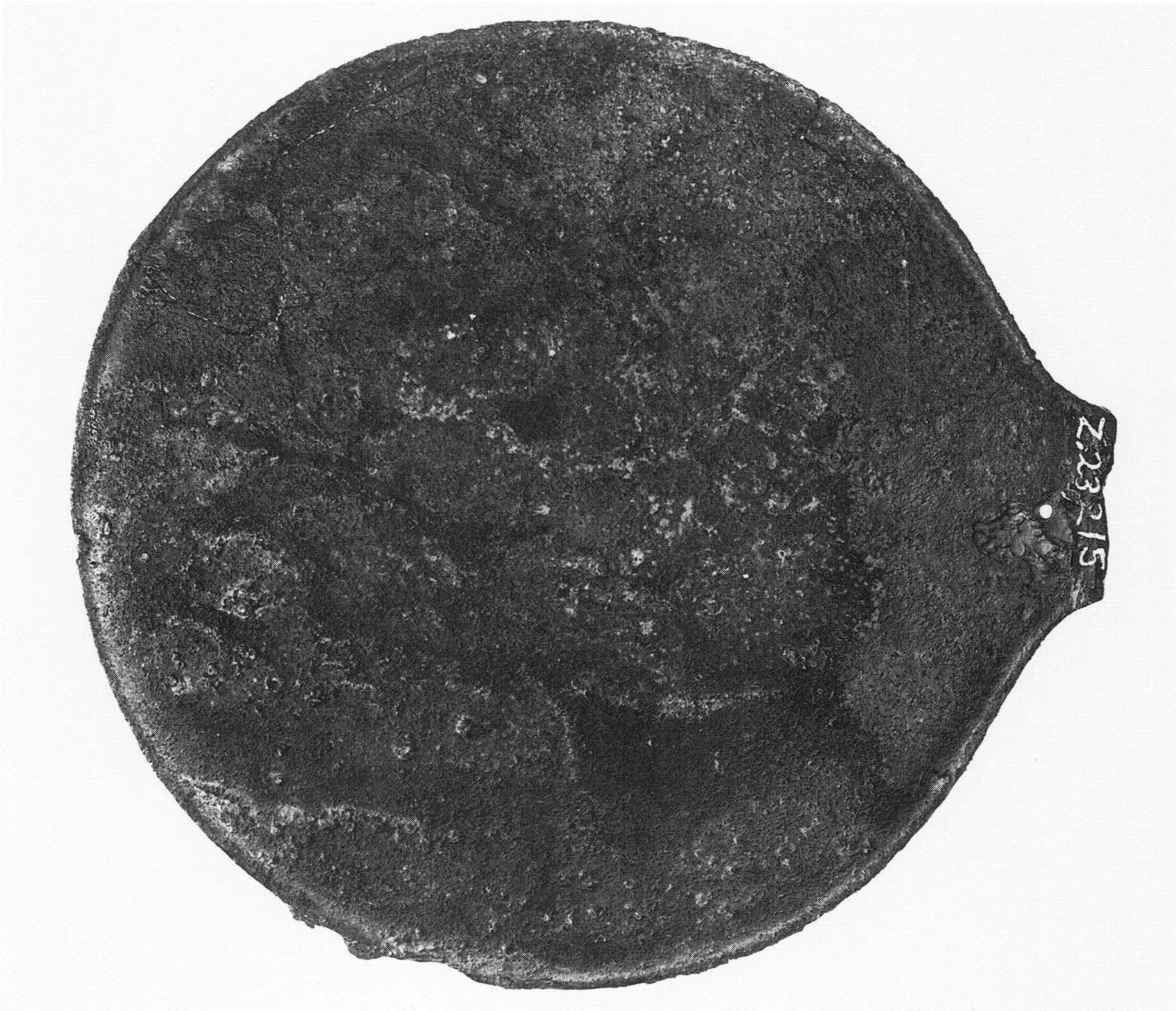

24d

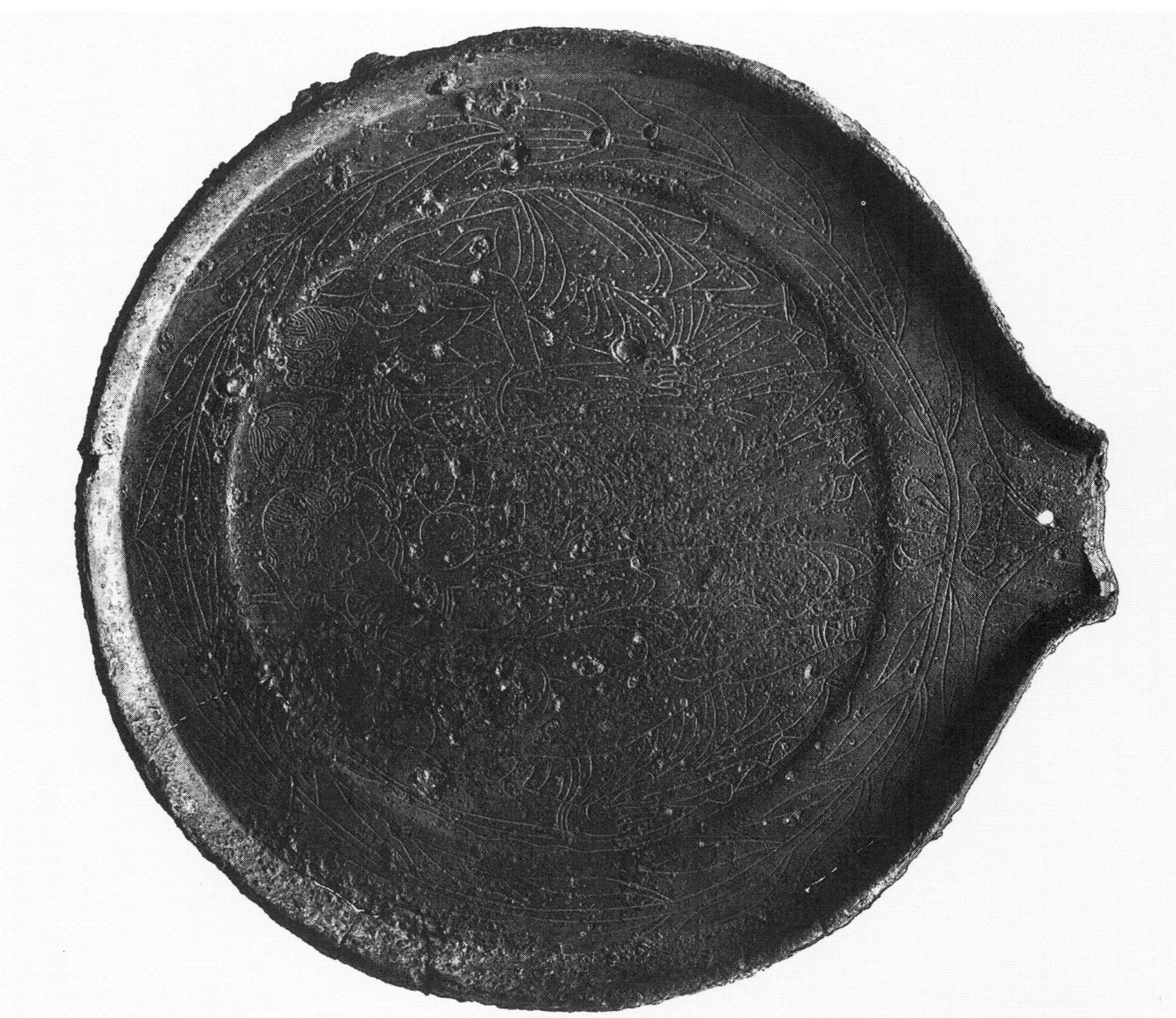

24b

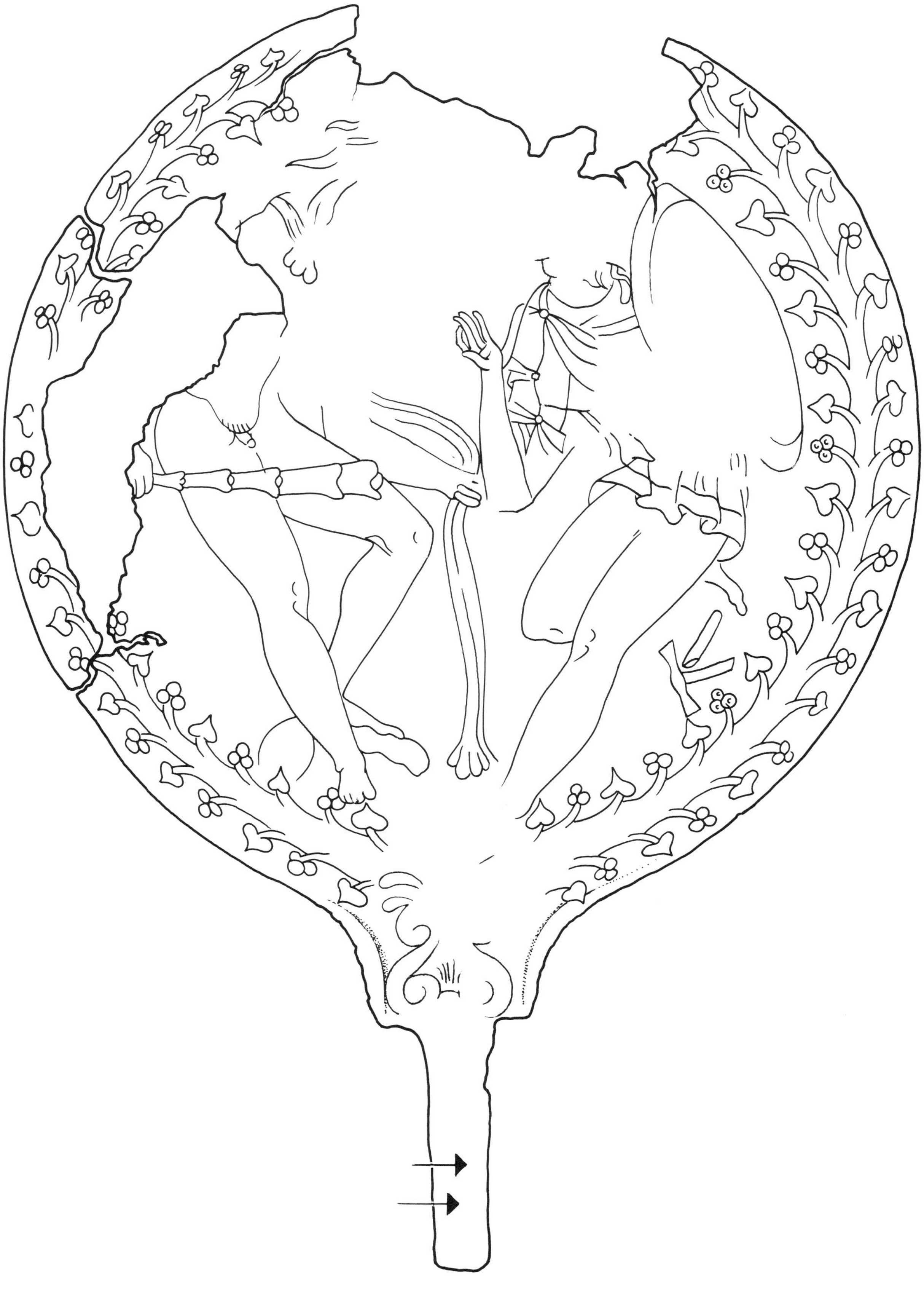

25a

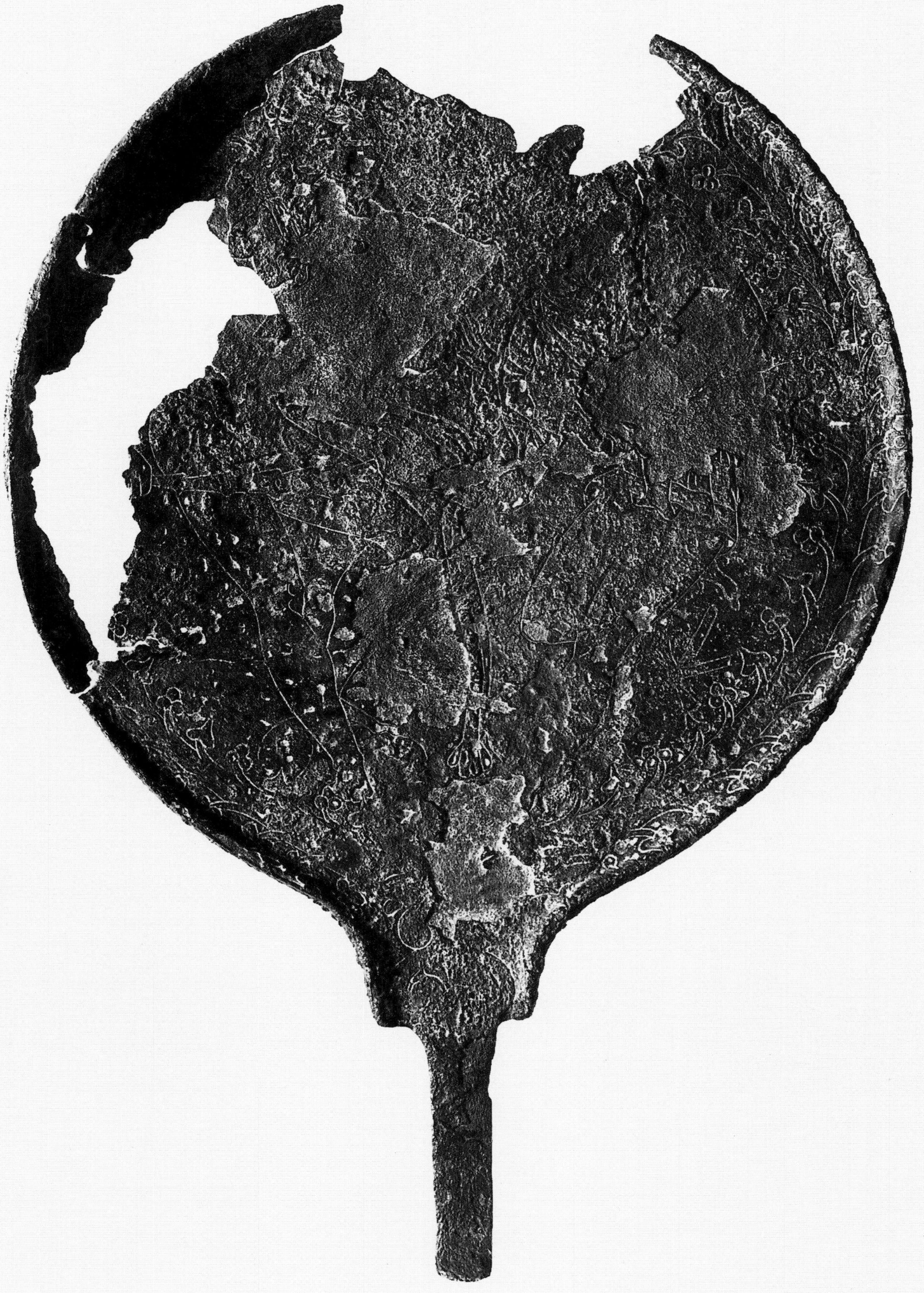

25b

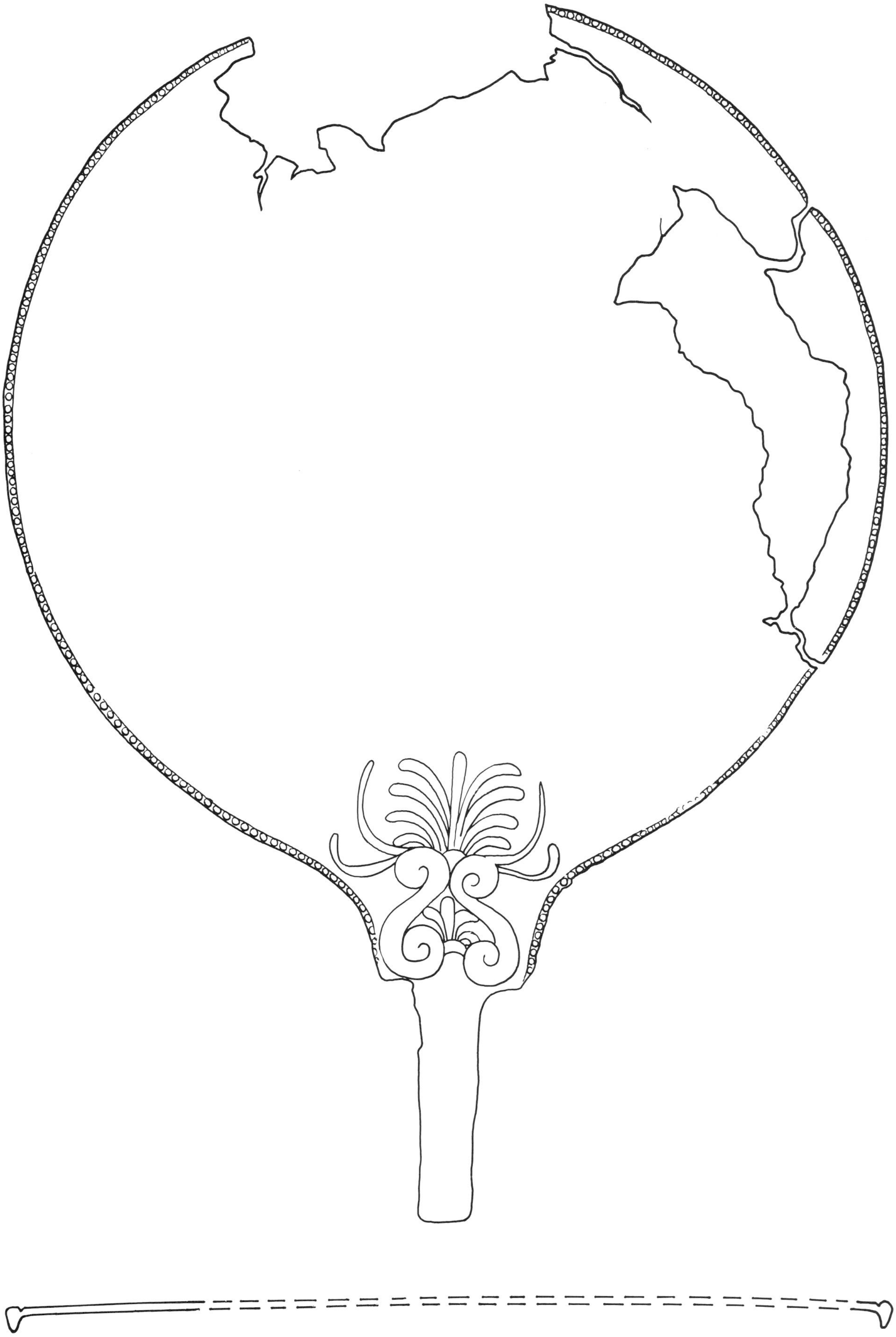

25c

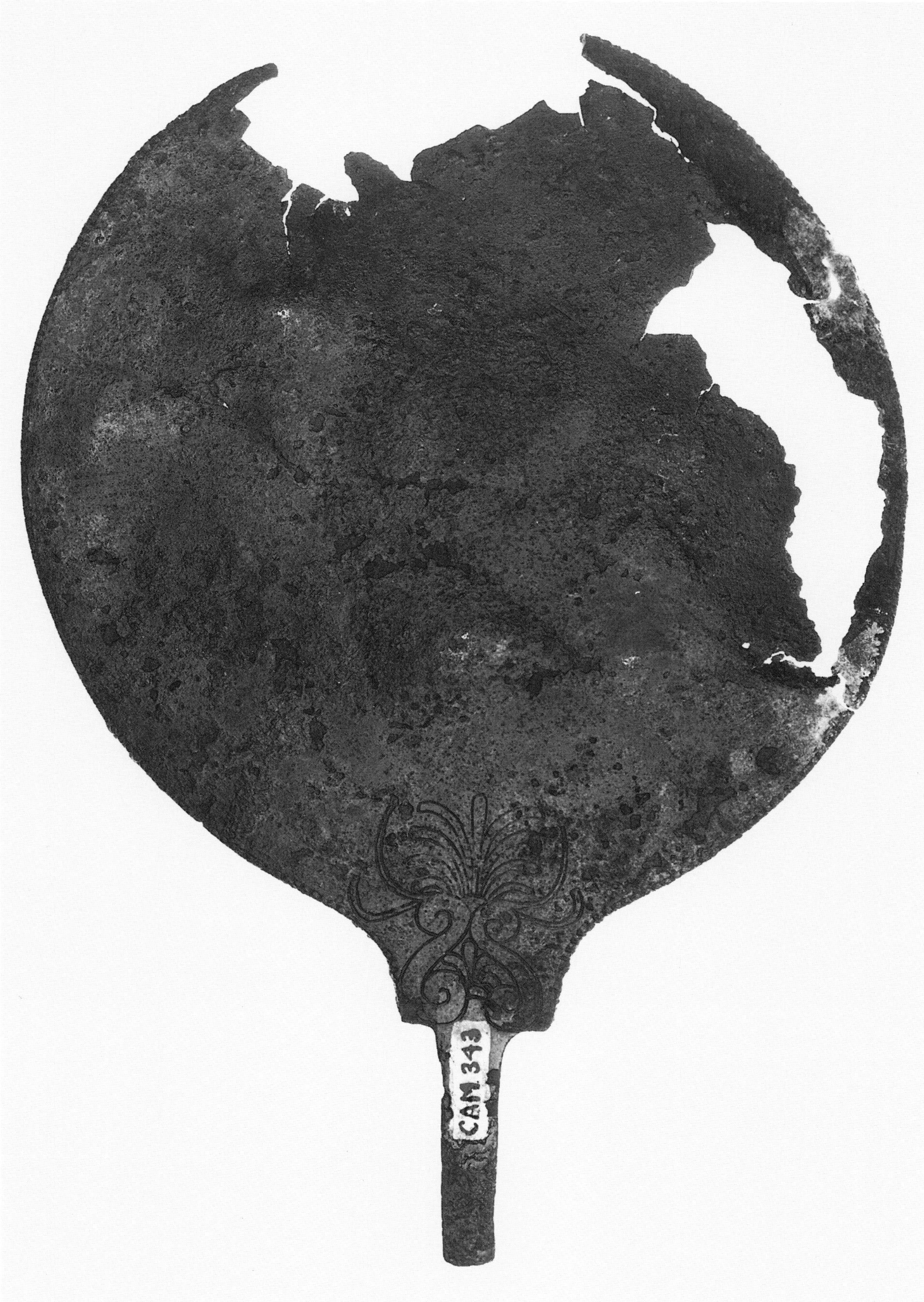

25d

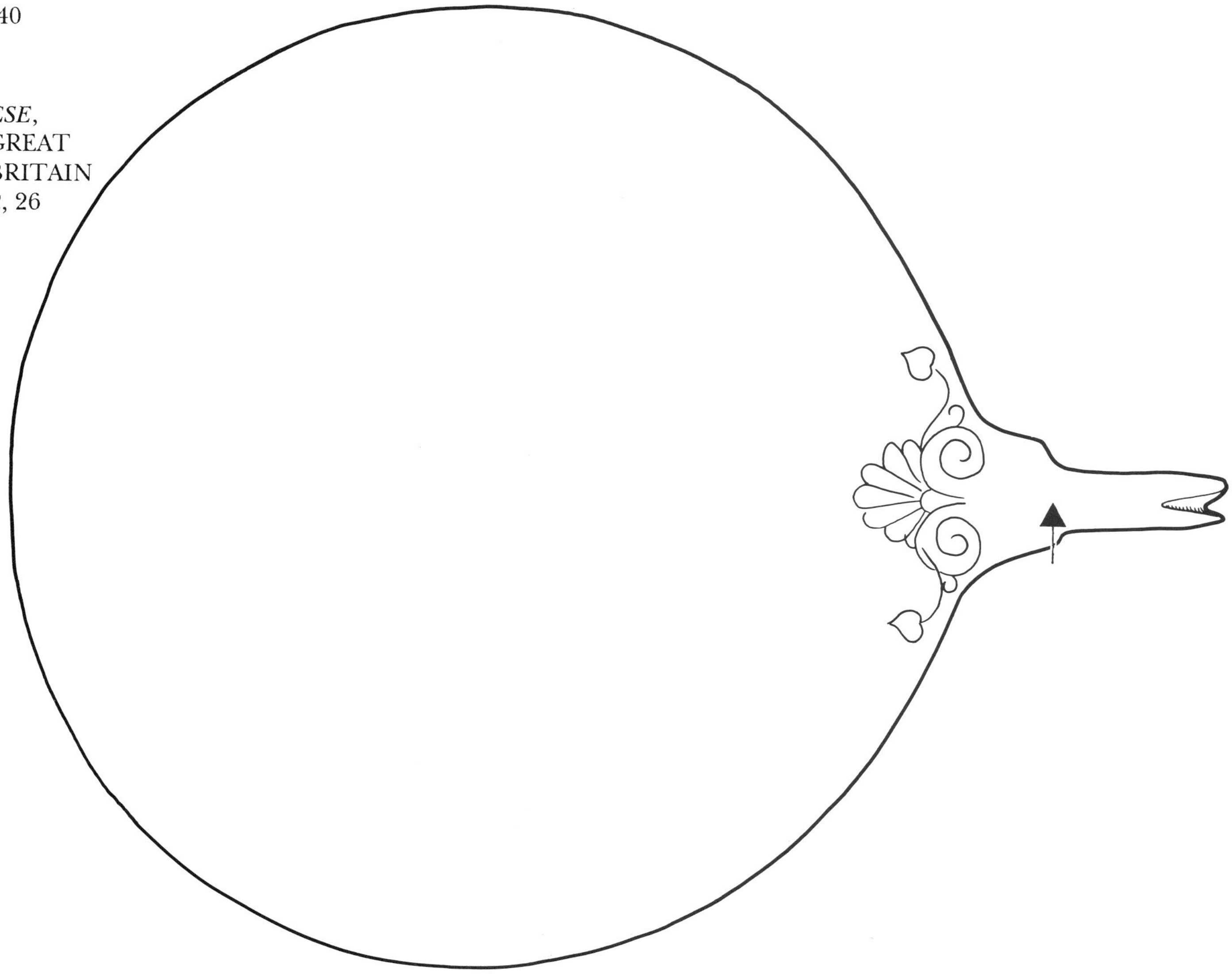

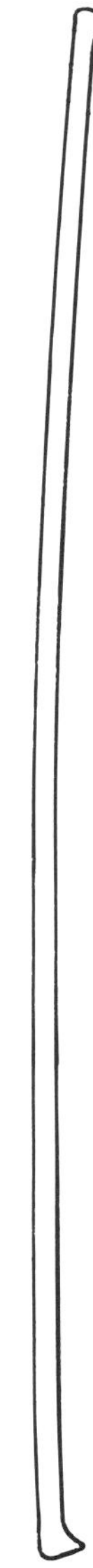

26c

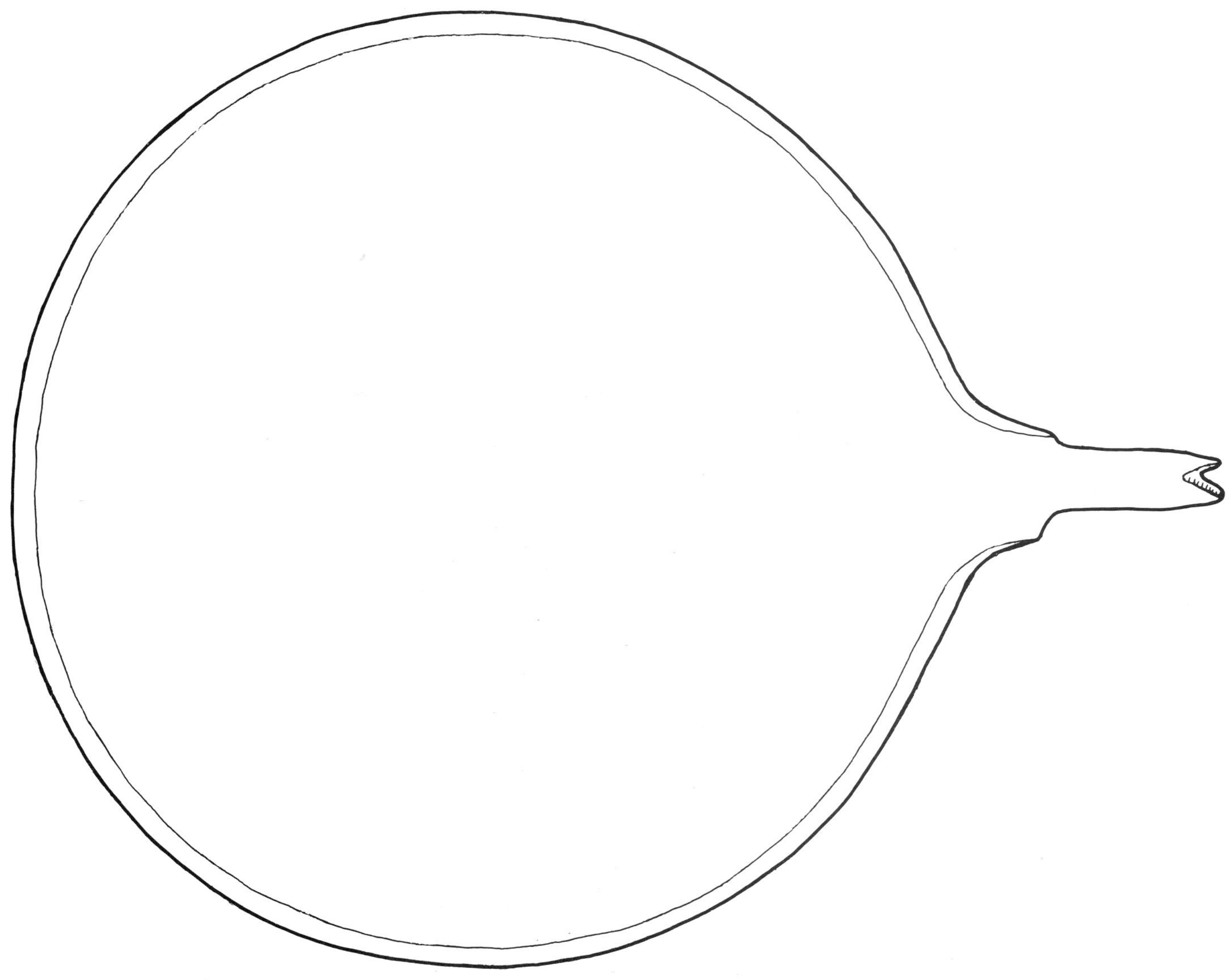

26a

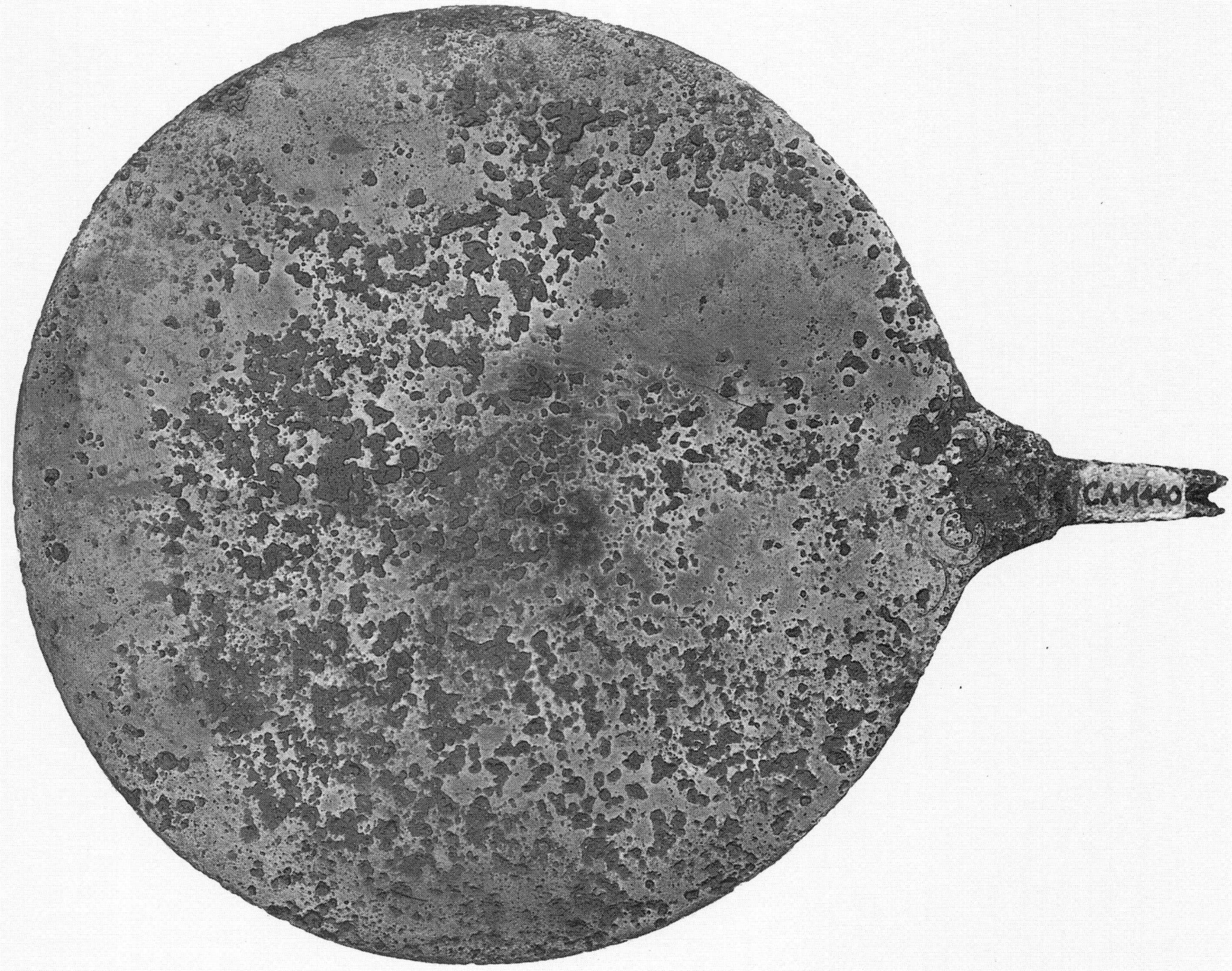

26d

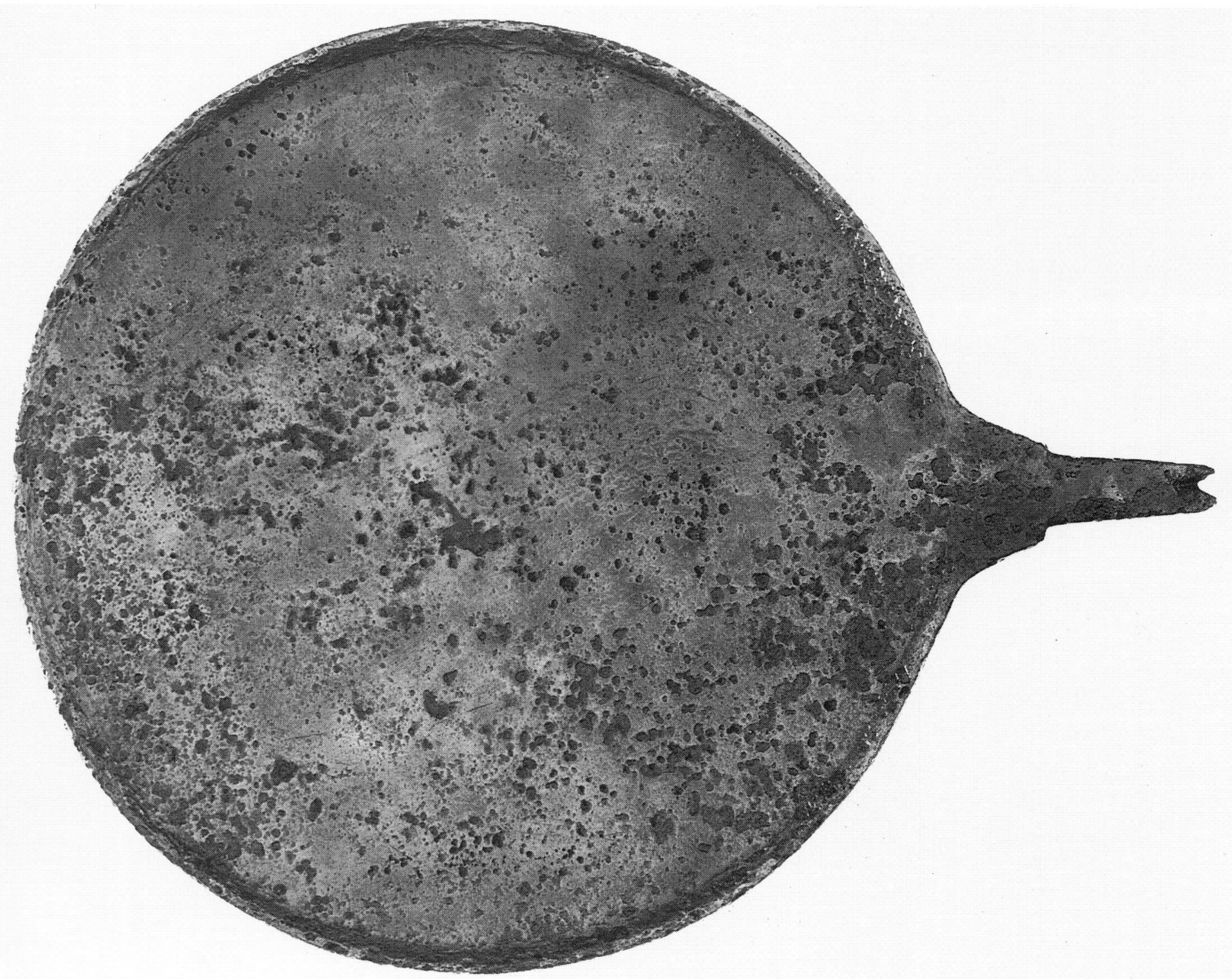

26b